Correlation of USA Daily Math Gra
to Common Core State Standards for Ma

D1066735

5.OA	Operations and Algebraic Thinking (Mondays)
5.OA.1 Use parentheses, brackets, or braces in numerical expressions.	p. 1 #3 p. 7 #3 p. 12 Brain Stretch p. 13 #2 p. 28 #1 p. 31 #1, 4 p. 34 #2, 4 p. 37 #2 p. _____ #1
5.OA.2 Write and interpret numerical expressions without evaluating.	p. 4 #2 p. 10 #1, 4 p. 19 #4 p. 22 #1, 3–4 p. 25 #3–4 p. 28 #3–4 p. 31 #3 p. 34 #1, 3 p. 37 #3 p. 43 #2 p. 46 #3 p. 52 #1, 3 p. 55 #2–3 p. 57 Brain Stretch p. 61 #2–3 p. 64 #2 p. 70 #2 p. 73 #1, 3 p. 76 #1, 3 p. 82 #2–3 p. 85 #2, 4
5.OA.3 Generate two numerical patterns using two given rules and graph the relationship.	p. 37 #1 p. 40 all p. 46 #1 p. 49 all p. 55 #1 p. 58 all p. 64 #1 p. 67 all p. 79 all p. 88 all
5.NBT	Numbers and Operations in Base 10 (Tuesdays)
5.NBT.1 Recognize place value in a multi-digit number.	p. 1 #1 p. 4 #1 p. 7 #3 p. 10 #1 p. 13 #1 p. 16 #3 p. 19 #1 p. 22 #2 p. 25 #1–2 p. 28 #1–2 p. 31 #1 p. 37 #1 p. 40 #2 p. 43 #2 p. 46 #1–2 p. 49 #1 p. 55 #1–2 p. 61 #1 p. 64 #1, 3 p. 70 #1 p. 73 #3 p. 76 #1 p. 79 #2 p. 82 #2 p. 88 #1
5.NBT.2 Explain patterns in number of zeroes when multiplying and decimal point placement when multiplying and dividing a decimal by a power of 10.	p. 7 #2 p. 10 #3 p. 13 #2 p. 16 #1–2 p. 22 #2 p. 31 #2 p. 40 #1 p. 49 #2 p. 52 #2 p. 73 #1 p. 82 #3 p. 85 #2
5.NBT.3 Read, write, and compare decimals to thousandths.	p. 1 #3 p. 4 #3 p. 16 #4 p. 19 #2–3 p. 25 #3–4 p. 31 #3 p. 34 #2 p. 37 #3 p. 40 #3, 5 p. 46 #3–4 p. 49 #3 p. 52 #5 p. 67 #2 p. 70 #2 p. 73 #2, 5 p. 85 #1
5.NBT.4 Round decimals.	p. 10 #2 p. 13 #3 p. 28 #3 p. 34 #1 p. 40 #4 p. 43 #1 p. 52 #1, 3 p. 55 #3 p. 58 #1–2 p. 61 #2 p. 67 #1 p. 76 #2 p. 79 #5 p. 88 #2
5.NBT.5 Multiply multi-digit whole numbers.	p. 22 #4 p. 31 #4 p. 58 #4 p. 67 #4 p. 72 Brain Stretch p. 73 #4 p. 82 #4 p. 85 #2, 4 p. 88 #3
5.NBT.6 Find whole-number quotients of whole numbers with up to four-digit dividends and two-digit divisors, using strategies.	p. 34 #4 p. 43 #3 p. 79 #4 p. 81 Brain Stretch p. 82 #1 p. 84 Brain Stretch
5.NBT.7 Add, subtract, multiply, and divide decimals to hundredths.	p. 1 #4 p. 4 #4 p. 7 #4 p. 10 #4 p. 13 #4 p. 19 #3–4 p. 22 #3 p. 25 #3 p. 28 #4 p. 31 #3 p. 33 Brain Stretch p. 34 #2–3 p. 37 #4 p. 40 #3 p. 43 #4 p. 46 #3 p. 49 #3 p. 51 Brain Stretch p. 52 #4 p. 55 #4 p. 58 #3 p. 63 Brain Stretch p. 64 #4 p. 67 #3 p. 70 #3–4 p. 72 Brain Stretch p. 75 Brain Stretch p. 76 #3 p. 79 #3 p. 82 #3 p. 85 #3 p. 88 #4
5.NF	Number and Operations (Wednesdays)
5.NF.1 Add and subtract fractions with unlike denominators, including mixed numbers.	p. 8 #2 p. 14 #1 p. 20 #1 p. 26 #2 p. 29 #1 p. 35 #1 p. 36 Brain Stretch p. 38 #1 p. 41 #1 p. 44 #2 p. 47 #1 p. 50 #2 p. 53 #1 p. 56 #2 p. 59 #2 p. 62 #2 p. 65 #1, 4 p. 69 Brain Stretch p. 71 #1, p. 80 #1 p. 83 #1
5.NF.2 Solve word problems involving addition and subtraction of fractions, including unlike denominators.	p. 2 #2 p. 5 #4 p. 8 #4 p. 9 Brain Stretch p. 11 #4 p. 20 #4 p. 27 Brain Stretch p. 29 #4 p. 32 #4 p. 44 #4 p. 39 Brain Stretch p. 77 #4

5.NF.3 Interpret a fraction as division and solve word problems involving whole numbers with fraction or mixed numbers answers.	p. 17 #3 p. 23 #4 p. 33 Brain Stretch p. 38 #4 p. 42 Brain Stretch p. 54 Brain Stretch
5.NF.4 Multiply a fraction or whole number by a fraction.	p. 17 #4 p. 32 #3 p. 35 #4 p. 38 #3 p. 44 #3 p. 47 #4 p. 54 Brain Stretch p. 56 #3 p. 62 #3–4 p. 65 #3 p. 74 #3 p. 80 #4 p. 83 #2 p. 86 #2 p. 89 #2
5.NF.5 Interpret multiplication as scaling (resizing).	p. 41 #3 p. 50 #3 p. 68 #2 p. 74 #2 p. 86 #1 p. 89 #1
5.NF.6 Solve real world problems involving multiplication of fractions and mixed numbers.	p. 18 Brain Stretch p. 24 Brain Stretch p. 41 #4 p. 53 #4 p. 56 #4 p. 68 #3–4 p. 74 #4 p. 77 #2 p. 80 #3 p. 83 #3–4
5.NF.7 Divide unit fractions by whole numbers and whole numbers by unit fractions.	p. 14 #3 p. 20 #3 p. 26 #4 p. 29 #3 p. 35 #2 p. 47 #3 p. 50 #4 p. 53 #3 p. 59 #4 p. 65 #2 p. 69 Brain Stretch p. 71 #3–4. p. 77 #3 p. 80 #2 p. 86 #4 p. 89 #3
5.G	**Geometry and Operations (Thursdays)**
5.G.1 Understand conventions and use a coordinate system.	p. 2 #1–3 p. 11 #1–3 p. 17 #1–3 p. 20 #1–3 p. 29 #1–3 p. 38 #1–3 p. 44 #1–2 p. 50 #1 p. 56 all p. 62 #1 p. 65 all p. 74 #a p. 83 #1–2
5.G.2 Graph and interpret points on a coordinate plane.	p. 2 #1–3 p. 11 #1–3 p. 17 #1–3 p. 20 #1–3 p. 29 #1–3 p. 38 #1–3 p. 44 #1–2 p. 50 #1–2 p. 56 all p. 62 #1, 3–4 p. 65 all p. 74 #a–b p. 83 #1–2
5.G.3 Understand that attributes of a category of 2-D figures belong to all subcategories of that category.	p. 14 #4 p. 32 #1 p. 41 #1 p. 53 #1 p. 59 #3 p. 68 #1 p. 77 #1 p. 80 #4 p. 86 #1 p. 89 all
5.G.4 Classify 2-D figures based on properties.	p. 14 #2 p. 35 #1 p. 47 #2 p. 71 #2 p. 89 all
5.MD	**Measurement and Data (Fridays)**
5.MD.1 Convert like measurement units within a given measurement system.	p. 3 #1, 3–4, Brain Stretch p. 6 #1–2 p. 9 #1–2 p. 15 #1–2, Brain Stretch p. 18 #1–3 p. 21 Brain Stretch p. 24 #1–2 p. 26 #1 p. 33 #1–2 p. 36 #1, Brain Stretch p. 39 #1–3 p. 45 #1, Brain Stretch p. 51 #1–3 p. 54 #1–2 p. 60 #1, Brain Stretch p. 63 #1 p. 69 #1, 4 p. 75 #1–3 p. 78 #1, Brain Stretch p. 81 Brain Stretch p. 84 #2 p. 87 #1
5.MD.2 Make a line plot to display measurements in fractions of a unit.	p. 48 #1–4 p. 57 #1–3 p. 72 #1–4 p. 90 #1–2
5.MD.3 Understand concepts of volume measurement.	p. 3 #2 p. 6 #3 p. 9 #3–4 p. 30 Brain Stretch
5.MD.4 Measure volumes by counting unit cubes.	p. 6 #4 p. 15 #4 p. 18 #4 p. 24 #4 p. 26 #4 p. 30 Brain Stretch p. 33 #3 p. 36 #3 p. 39 #4 p. 51 #4 p. 60 #2
5.MD.5 Relate volume to multiplication and addition and solve problems related to volume.	p. 21 Brain Stretch p. 36 #4 p. 45 #4 p. 48 Brain Stretch p. 54 #4 p. 60 #2, 4 p. 63 #2–3 p. 69 #2 p. 75 #4 p. 78 #2–3 p. 84 #1 p. 87 #3–4

MONDAY — Operations and Algebraic Thinking

1 Which property does this equation show?

$3 + 7 = 7 + 3$

A. commutative

B. associative

C. distributive

2 $2 + 80 - 31 - 7 + 3 =$

3 Add parentheses to make the equation true.

$21 + 1 \times 2 = 44$

4 What is the pattern rule?

9, 18, 27, 36, 45, 54

TUESDAY — Operations in Base Ten

1 Fill in the blank to compare the numbers.

400 is _____ times larger than 40

2 $32 \times 10 =$

3 Compare the decimals using <, >, or =.

0.398 ☐ 0.331

4 Add. Use words, pictures, or equations to show your work.

$0.3 + 0.5 + 0.8 =$

WEDNESDAY Fractions

1 Write two equivalent fractions.

$\frac{2}{4}$

2 Mita rode her bike $3\frac{1}{3}$ miles to her grandmother's house. Then she rode $1\frac{1}{2}$ miles to the arena. Choose the best estimate of the total distance Mita traveled on her bike.

   A. about $4\frac{1}{2}$ miles

   B. about 6 miles

   C. about $6\frac{1}{2}$ cups

3 Compare the fraction to the number using <, >, or =.

$\frac{30}{7}$ ☐ 5

4 Write the mixed number as an improper fraction.

$4\frac{6}{9}$

THURSDAY Geometry

1 Write the coordinate pair for each item on the coordinate plane.

★ _____   ✚ _____   ◆ _____

 _____   ⚓ _____   🖱 _____

🦌 _____   ⬆ _____   ⛷ _____

2 Draw a circle at (4, 9).

3 Draw a triangle at (0, 5).

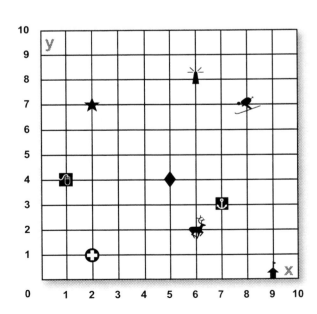

Chalkboard Publishing © 2012

1 7 m = _____ cm

2 Which 3D shapes would you use to measure the volume of a box?

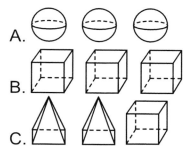

A.

B.

C.

3 Which container will hold more liquid: a 3-liter container or a 3,000-milliliter container?

4 The time is 10:34 a.m.
What time will it be in 285 minutes?

BRAIN STRETCH

A box of cookies weighs 520 grams. Amanda brought three boxes to the class party. What is the combined weight of the boxes in kilograms?

MONDAY — Operations and Algebraic Thinking

1 Number the operations according to the order of operations:

___ multiplication and division

___ brackets and exponents

___ addition and subtraction

3 21 + 8 ÷ 4 + 5 =

2 Compare the expressions without calculating. Use <, >, or =.

22 + 11 + 33 ▢ 11 + 22 + 33

4 What is the pattern rule?

6, 12, 24, 48, 96

TUESDAY — Operations in Base Ten

1 6 hundreds = _____ ones

2 80 × 100 =

3 Compare the decimals using <, >, or =.

0.673 ▢ 0.67

4 Subtract. Use words, pictures, or equations to show your work.

10 − 7.4 =

WEDNESDAY Fractions

1 Write two equivalent fractions.

$\dfrac{1}{8}$

2 Compare the fractions using <, >, or =.

$\dfrac{6}{8}$ ☐ $\dfrac{1}{8}$

3 Write the improper fraction as a mixed number.

$\dfrac{15}{8}$

4 Samir combined $2\frac{1}{3}$ cups of oats, $\frac{2}{4}$ cup of raisins, and $5\frac{1}{3}$ cups of nuts to make some trail mix. How much trail mix did he make? Use a model or an equation to show your work.

THURSDAY Geometry

1 Circle all the polygons.

2 Color the regular polygons green.

3 Color the irregular polygons blue.

4 Choose one shape that is not a polygon. Describe how you could change it into a polygon.

1 90 lb = _____ oz

2 A rug is $2\frac{1}{2}$ yards long. How long is the rug in inches?

3 Which cube would you use to measure the volume of a box?

A.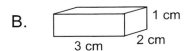

B.

C.

4 What is the volume of this shape in cubic units? _____

BRAIN STRETCH

The city of Oakdale wants students to walk in the annual parade and throw out candy to the crowds. The city wants one student with the first float, two students with the second float, three students with the third float, four students with the fourth float, and so on. There are 10 floats in the parade. If the pattern continues, how many students in total will walk in the parade?

MONDAY — Operations and Algebraic Thinking

1 Which property does this equation show?

$8 \times 60 = 60 \times 8$

A. distributive

B. associative

C. commutative

2 $50 + 4 \times 2 \times 2 =$

3 Add parentheses to make the equation true.

$7 \times 7 - 2 + 10 = 45$

4 What is the pattern rule?

1,000   950   900   850   800

TUESDAY — Operations in Base Ten

1 Write the number in expanded form.

$238,675 =$

2 a) Complete the pattern.

$3 \times 1 = \underline{\hspace{1cm}}$

$3 \times 10 = \underline{\hspace{1cm}}$

$3 \times 100 = \underline{\hspace{1cm}}$

$3 \times 1,000 = \underline{\hspace{1cm}}$

b) Find $3 \times 100,000$ without multiplying.

3 Fill in the blank to compare the numbers.

0.5 is _____ times smaller than 5

4 Subtract. Use words, pictures, or equations to show your work.

$20.5 - 5.4 =$

WEDNESDAY Fractions

1 Write the improper fraction as a mixed number.

$\dfrac{5}{3}$

2 Add.

$\dfrac{5}{8} + \dfrac{3}{4} =$

3 Compare the fraction to the number using <, >, or =.

$\dfrac{3}{4}$ ☐ 1

4 Cameron lives $7\frac{1}{4}$ miles from the movie theater. Tonya lives $3\frac{2}{5}$ miles from the theater. How much farther is the movie theater from Cameron's house than Tonya's house? Use a model or an equation to show your work.

THURSDAY Geometry

1 Classify the pair of lines

A. intersecting

B. perpendicular

C. parallel

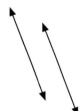

2 Which shapes are polygons?

A.   B.   C.   D.

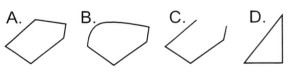

3 Draw an irregular polygon with 6 sides.

4 How many lines of symmetry are there?

P _____

1 4 quarts = _____ pints

2 Olivia was in a track and field competition. In the long jump event she jumped 4.45 meters. How many centimeters did she jump?

3 A box has a volume of eighteen cubic meters. Which is the correct way to write the volume?

A. 18 m   B. 18 m²   C. 18 m³

4 Thomas put 30 unit cubes in a box like this:

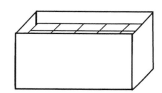

He says the volume of the box is 30 cubic units. Is this correct? Why or why not?

BRAIN STRETCH

Helen made 6 pizzas and cut each of them into eighths.
If she served 3¼ pizzas, how many slices of pizza did Helen have left over?

MONDAY — Operations and Algebraic Thinking

1 Which expression is **not** equal to 5 × 5?

A. 5 + 5 + 5 + 5 + 5
B. 4 × 5 + 5
C. 2 × 5
D. 5 + 4 × 5

2 13 + 7 − 8 ÷ 4 =

3 Complete the function table.
Rule: Output = Input × 12

Input	Output
3	
6	
9	
12	
15	

4 Write an algebraic expression for this description:

20 more than 17

TUESDAY — Operations in Base Ten

1 _____ hundreds = 40 tens

2 To which place value was the number rounded?

67,874 → 70,000

_____

3 a) Complete the pattern.

50 ÷ 10 = _____
500 ÷ 10 = _____
5,000 ÷ 10 = _____
50,000 ÷ 10 = _____

b) Find 90,000 ÷ 10 without dividing.

4 Add. Use words, pictures, or equations to show your work.

8.01 + 3.30 =

WEDNESDAY — Fractions

1 Write two equivalent fractions.

$$\frac{5}{10}$$

2 Write the mixed number as an improper fraction.

$$6\frac{2}{3}$$

3 Compare the quantities using <, >, or =.

$$3 \times \frac{3}{4} \boxed{} 3$$

4 Sebastian and his family picked $4\frac{1}{3}$ bushels of red apples and $2\frac{1}{8}$ bushels of green apples. How many bushels of apples did the family pick in all? Use a model or an equation to show your work.

THURSDAY — Geometry

1 Write the coordinate pair for each point.

A _____ B _____ C _____

D _____ E _____ F _____

G _____ H _____ I _____

2 Draw a point at (2, 2).

3 Draw a point at (0, 3).

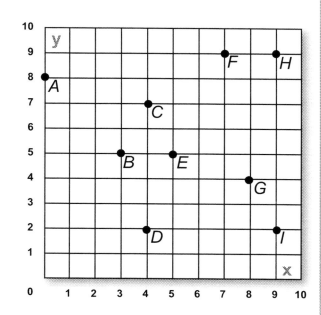

Jane asked all the students in her class how many times they visited the library to get research materials for a class project. This is the data she collected:

Number of Visits to the Library

1 2 0 2 4 2 0 2 2 2 4 2 3 3 3 4 4 4 3 4 4 2 4 4

1 Use Jane's data to complete the line plot.

<----|-------|-------|-------|-------|---->

2 What can you conclude from the line plot?

_____

_____

BRAIN STRETCH

Noor invited 5 people to a fundraiser. Everyone shook everyone else's hand to say hello. How many handshakes were there? Show your work.

MONDAY — Operations and Algebraic Thinking

1 Which property does this equation show?

$4 + (10 + 2) = (4 + 10) + 2$

A. commutative
B. associative
C. distributive

2 $56 \times (2 + 4 - 1) =$

3 What is the pattern rule?

67, 73, 77, 83, 87

4 Circle the numbers that are **not** a multiple of 12.

144   24   56   36   96

TUESDAY — Operations in Base Ten

1 What is the value of 2 in each number?

742 _____

724 _____

2 Multiply. Then write the answer using an exponent.

$10 \times 10 \times 10 =$

3 Round 9.322 to the nearest hundredth.

4 Add. Use words, pictures, or equations to show your work.

$50.24 + 10.5 =$

Chalkboard Publishing © 2012

WEDNESDAY — Fractions

1 Subtract, then simplify your answer.

$$\frac{5}{6} - \frac{1}{3} =$$

2 Compare the fraction to the number using <, >, or =.

$$\frac{21}{2} \boxed{} 10$$

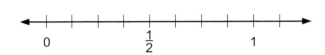

3 How much fudge will each person get if 5 people share $\frac{1}{2}$ lb of fudge equally? Use a model to show your work.

4 Plot $\frac{6}{8}$ on the number line. Is it closest to 0, $\frac{1}{2}$, or 1?

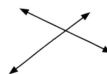

THURSDAY — Geometry

1 Classify the pair of lines.

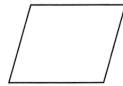

A. intersecting

B. perpendicular

C. parallel

2 Draw and name 2 polygons that have more than 4 sides.

3 What do we call a shape that has four sides and no parallel sides?

4 Choose all the words that describe this shape.

A. parallelogram

B. rhombus

C. quadrilateral

D. square

1 100 years = _____ months

2 Sharon walked 2 miles to the beach. How many yards did she walk?

3 What are the perimeter and area of a tabletop that is 56 cm wide and 80 cm long?

4 What is the volume of this shape in cubic units?

_____

BRAIN STRETCH

Adele has to be at school by 8:30 a.m. It takes her 15 minutes to get dressed, 10 minutes to eat, and 25 minutes to walk to school. What time should she get up?

Operations and Algebraic Thinking

1 (13 − 5 + 83) × 2 =

2 Complete the function table.
Rule: Output = Input ÷ 8 + 7

Input	Output
8	
88	
32	
64	
16	

3 Classify the numbers as prime (P) or composite (C).

A. 39 _____

B. 87 _____

4 Add parentheses to make the equation true.

2 × 2 + 12 ÷ 4 + 3 = 4

TUESDAY Operations in Base Ten

1 a) 34 × 1 = _____

b) 34 × 10 = _____

c) 34 × 100 = _____

d) What do you notice about the number of zeroes in the product when you multiply by powers of 10?

2 Multiply. Then write the answer using an exponent.

10 × 10 × 10 × 10 × 10 =

3 How many tenths are in 6.7?

4 Compare using <, >, or =.

0.450 ☐ 0.451

WEDNESDAY — Fractions

1 Write two equivalent fractions.

$\dfrac{4}{7}$

2 Compare the quantities using <, >, or =.

$2\dfrac{1}{4}$ ☐ $\dfrac{12}{4}$

3 Use a model to find $5 \times \dfrac{5}{6}$.

4 If 7 people want to share a 40-pound sack of flour equally by weight, how many pounds of flour should each person get? Between what two whole numbers does the answer lie? Use a model or an equation to show your work.

THURSDAY — Geometry

1 Write the coordinate pair for each point.

A _____ B _____ C _____

D _____ E _____ F _____

G _____ H _____ I _____

2 Draw a point at (0, 0).

3 Draw a point at (8, 1).

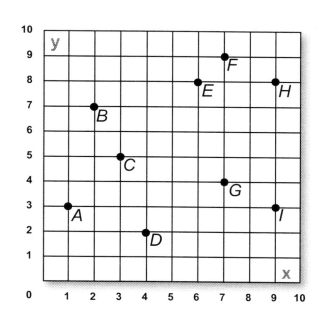

Measurement and Data

1 23 pounds = _____ ounces

2 How many pennies are in $100?

3 Connor trained for the track meet by running 2.4 km every day. How many meters did he run every week?

4 What is the volume of this shape in cubic units?

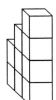

_____

BRAIN STRETCH

Maggie earns $8 per hour for the first 8 hours that she works each day and $1\frac{1}{2}$ times this amount for each hour that she works over 8 hours. How much did Maggie earn if she worked 9 hours last Saturday?

Operations and Algebraic Thinking

1 Which property does this equation show?

$(3 \times 7) \times 12 = 3 \times (7 \times 12)$

A. distributive
B. associative
C. commutative

3 $(6 \times 30) - [6 + (80 \div 8)] =$

2 Complete the function table.
Rule: Output = Input × 9 + 5

Input	Output
7	
6	
10	
11	
4	

4 Compare the expressions using <, >, or =.

$60 + 2 \times 3 \boxed{} 11 \times (3 + 3)$

Operations in Base Ten

1 9 hundreds = _____ ones

3 Complete the sentence using <, >, or =.

$9.3 - 0.2 \boxed{} 9.1$

2 Write the number in expanded form.

$186.302 =$

4 Multiply. Use words, pictures, or equations to show your work.

$5.5 \times 10 =$

 Chalkboard Publishing © 2012

WEDNESDAY · Fractions

1 Add.

$$\frac{2}{3} + \frac{1}{9} =$$

2 Compare the quantities using <, >, or =.

$$2\frac{1}{4} \ \boxed{\phantom{<}} \ \frac{12}{4}$$

3 Use a model to divide $\frac{1}{9} \div 2$.

4 Charlie had $4\frac{1}{2}$ cups of tomato sauce in a large can. He used three quarters of a cup to make spaghetti sauce.
How much tomato sauce is left in the large can?

THURSDAY · Geometry

1 Plot the points on the coordinate plane.

A (2, 0)	E (5, 7)
B (7, 1)	F (1, 3)
C (6, 2)	G (9, 3)
D (3, 4)	H (8, 6)

2 Which point is closest to the origin?

_____

3 Which point is closest to (5, 5)?

_____

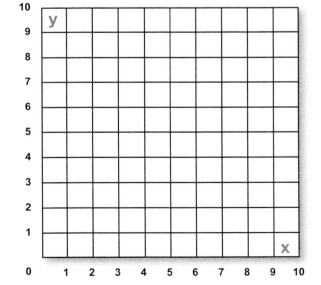

This double bar graph shows how many tulips of each color Chris and Sophie planted in their gardens.

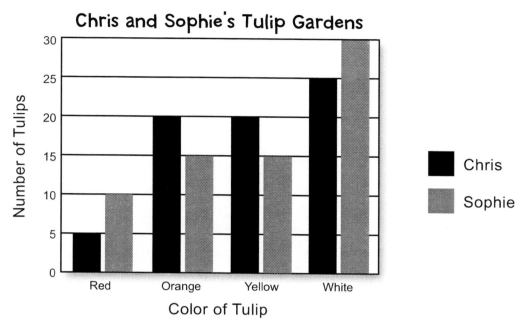

Chris and Sophie's Tulip Gardens

1 How many red tulips did Sophie plant? _____

2 How many fewer white tulips did Chris plant than Sophie? _____

3 Who planted more orange and yellow tulips? _____

4 Who planted the most tulips altogether, Chris or Sophie? Estimate by looking at the graph. Then add the totals to check your estimate.

_____

BRAIN STRETCH

A hotel has a pool in the shape of a rectangle. The pool is 8 m long, 3 m wide, and 2,000 cm deep. What is the volume of the pool in cubic meters?

1 Which expression is equal to 6 × 42?

A. 6 × (42 − 2)
B. 6 × (6 + 42)
C. 6 × (40 − 2)
D. 6 × (40 + 2)

3 Compare the expressions without calculating. Use <, >, or =.

(4 × 4) × 2 ☐ 4 × (4 × 2)

2 Find the rule and complete the function table.

_____

Input	Output
5	55
6	
7	77
8	88
9	

4 Write an algebraic expression for this description:

75 reduced by 25

1 80 thousands = _____ hundreds

2 $16 \times 10^3 =$

3 Subtract. Use words, pictures, or equations to show your work.

0.50 − 0.44

4
$$\begin{array}{r} 75 \\ \times\ 36 \\ \hline \end{array}$$

1 Write two equivalent fractions.

$$\frac{10}{22}$$

2 Write the mixed number as an improper fraction.

$$5\frac{1}{8}$$

3 Compare the fraction to the number using <, >, or =.

$$\frac{14}{2} \boxed{} \; 7$$

4 At a party, 12 people shared 2 apple pies equally. What fraction of pie did each person get? Show your work using pictures, words, and numbers.

THURSDAY — Geometry

1 Count the number of sides in each shape.

A _____ sides   B _____ sides   C _____ sides   D _____ sides

E _____ sides   F _____ sides   G _____ sides   H _____ sides

I _____ sides

J _____ sides

2 Sort the shapes. The first one is done for you.

Property	Shapes
Quadrilaterals	
Not Quadrilaterals	A

1 3 quarter hours = _____ minutes

2 Brian rides his bike to and from school each day. The distance from his house to the school is 1.8 km. How many meters does Brian ride each day? How many meters does he ride in five days?

3 Calculate the perimeter of the rectangle.

12.2 yards

5.1 yards

4 What is the volume of this shape in cubic units?

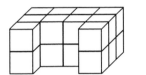

_____

BRAIN STRETCH

Dave has $90. A red sweater costs $52 and a blue sweater costs $\frac{5}{6}$ as much. Does Dave have enough money to buy both sweaters? Show your work.

MONDAY — Operations and Algebraic Thinking

1 Which property does this equation show?

$3 \times (4 + 2) = 3 \times 4 + 3 \times 2$

A. associative
B. distributive
C. commutative

2 $(35 + 19) - (4 - 3) \times 13 =$

3 How many times larger or smaller? Compare the expressions without calculating.

$9 \times (730 - 144)$ is _____ times

_____ than $(730 - 144) \times 3$

4 Write an algebraic expression for this description:

add 12 and 8, then divide by 2

TUESDAY — Operations in Base Ten

1 _____ tens = 400 ones

2 Each digit is 4, but the value is different.

a) How does the 4 that the arrow points to compare in value to the 4 to its left?

b) How does it compare in value to the 4 to its right?

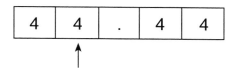

3 Complete the sentence using <, >, or =.

$5.3 + 2.3$ ☐ 8

4 Write the number in expanded form.

$872.549 =$

WEDNESDAY — Fractions

1 Write the improper fraction as a mixed number.

$$\frac{24}{9}$$

2 Add, then simplify your answer.

$$\frac{2}{12} + \frac{4}{5} =$$

3 Compare the quantities using <, >, or =.

$$4\frac{3}{4} \;\boxed{}\; \frac{20}{4}$$

4 How many $\frac{1}{3}$-cup servings are in 3 cups of yogurt? Use a model and show your reasoning.

THURSDAY — Geometry

1 Classify the pair of lines

A. intersecting

B. perpendicular

C. parallel

2 How are a square and a parallelogram alike?

3 Choose all the words that describe the shape.

A. right

B. equilateral

C. triangle

D. scalene

4 Are these shapes congruent or similar?

1 17 feet = _____ inches

2 Jerry can type 25 words per minute. At this rate, how many words can Jerry type in 4.5 minutes?

3 Which room has a larger area and perimeter? Justify your answer.

Room A   12 ft. by 8 ft.

Room B   9 ft. by 9 ft.

4 a) What is the volume of this shape in cubic units?

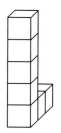

_____

b) What would be the total volume of 4 shapes like this one?

_____

BRAIN STRETCH

Sophia used $6\frac{1}{3}$ cups of pineapple juice, $2\frac{1}{2}$ cups of ginger ale, and $1\frac{1}{2}$ cups of orange juice to make some fruit punch. How many pints of punch did she make?

1 $(50 \times 4) + (99 \div 9) + 12 =$

2 Complete the function table.
Rule: Output = Input ÷ 5 × 10

Input	Output
45	
60	
5	
15	
30	

3 Write an algebraic expression for this description:

add 6 and 5, then multiply by 3

4 Which expression is **not** equal to 32?

A. 4 × 4 × 2

B. 4 × 8

C. 2 × 4 × 4

D. 8 × 2

1 10 hundred thousands

= _____ ten thousands

2 For the number 3333, the 3 in the tens place represents 30.

a) The 3 in the hundreds place represents_____.
b) The 3 in the thousands place represents _____.
c) The 3 in the hundreds place is _____ times as much as the 3 in the tens place.

3 Round 672,725.9 to the nearest ten thousand.

4 Divide. Use words, pictures, or equations to show your work.

$35.5 \div 5 =$

WEDNESDAY Fractions

1 Subtract.

$$\frac{8}{9} - \frac{3}{5} =$$

2 Compare the fraction to the number using <, >, or =.

$$\frac{7}{6} \boxed{} 2$$

3 Use a model to divide $\frac{1}{5} \div 4$.

4 Ashley used $6\frac{6}{8}$ inches of red ribbon and $7\frac{1}{4}$ inches of white ribbon to wrap a present. How many inches of ribbon did Ashley use altogether?

THURSDAY Geometry

1 Plot the points on the coordinate plane.

A (2, 2)   E (6, 6)
B (4, 5)   F (0, 9)
C (7, 9)   G (4, 4)
D (1, 3)   H (8, 8)

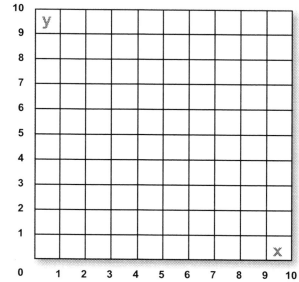

2 Which points have the same x-coordinate? _____
Which points have the same y-coordinate? _____

3 Draw a line between points A and F.
Which other points are on the line? _____

This pictograph shows the amount of ice cream sold at the Ice Hut on the weekend.

Ice Cream Sold (in pints)

Chocolate	🍦🍦🍦🍦🍦
Chocolate Chip	🍦🍦🍦🍦🍦🍦🍦🍦🍦
Strawberry	🍦🍦🍦🍦🍦🍦🍦🍦🍦🍦
Vanilla	🍦🍦🍦🍦🍦🍦🍦

Key: 🍦 = 1 pint

1 How many pints of each flavor were sold?

_____

2 If the Ice Hut had 2 gallons of each flavor at the beginning of the weekend, how many pints of each flavor are left on Monday morning?

_____

BRAIN STRETCH

At Super Button Supplies, workers pack 6 buttons into cubic boxes with sides 1 inch. They pack the small cubic boxes into larger cubic boxes with sides 8 inches.

a) How many small boxes will fit in each larger box?
b) How many buttons are in each larger box?

Operations and Algebraic Thinking

1 $[(12 \times 4) + 2] \div 2 =$

2 Find the rule and complete the function table.

Input	Output
4	400
7	700
8	
10	
	1,100

3 Write an algebraic expression for this description:

add 8 and 56, then divide by 4

4 Add parentheses to make the equation true.

$16 + 4 \times 8 - 1 = 140$

TUESDAY Operations in Base Ten

1 2 ten thousands = _____ ones

2 $51 \times 10^5 =$

3 Complete the sentence using <, >, or =.

4.7 − 0.5 ☐ 4.5

4
```
   96
 × 61
```

WEDNESDAY — Fractions

1 Write two equivalent fractions.

$\frac{3}{5}$

2 Write the improper fraction as a mixed number.

$\frac{14}{13}$

3 Each square in the grid has sides that measure $\frac{1}{4}$ ft. What is the area of the rectangle in feet?

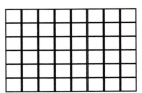

4 Mrs. Mudhar needs at least 16 yards of fabric to make costumes for a school play. She has $4\frac{1}{2}$ yards of fabric at home and she buys $10\frac{1}{3}$ yards of the same fabric at the store. Does she have enough fabric for the costumes or does she need to buy more? Justify your answer.

THURSDAY — Geometry

1 Which shape is **not** a quadrilateral?

A. octagon

B. rhombus

C. square

2 Classify the triangle. Circle all the descriptions that apply.

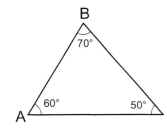

A. acute

B. isosceles

C. obtuse

D. scalene

E. right

3 How many lines of symmetry are there?

X _____

4 Are these shapes congruent or similar?

1 2.3 L = _____ mL

2 Katelyn's bedroom is 4.7 m by 5 m. What is the area of her room in centimeters?

3 Which shape has the largest volume?

A.

B.

C.

4 How much carpet would you need to cover the floor in a room that is 14 feet by 5 feet?

BRAIN STRETCH

Carlos went shopping for cereal. Honey Hoops cereal costs $1.50/500 g and Fruity Flakes cereal costs $0.55/100 g. Which cereal is the better buy? Show your reasoning.

MONDAY — Operations and Algebraic Thinking

1 Which expression is equal to $4 \times (60 - 5)$?

A. 4×55

B. 24×20

C. 240×5

D. 20×60

2 $60 - (45 \div 9) =$

3 Write an algebraic expression for this description:

subtract 5 from 11, then add 2

4 Add parentheses to make the equation true.

$16 + 3 \times 4 - 2 + 5 = 21$

TUESDAY — Operations in Base Ten

1 Round 12.846 to the nearest hundredth.

2 Complete the sentence using <, >, or =.

$2.4 + 0.5 \boxed{} 3.9$

3 Multiply. Use words, pictures, or equations to show your work.

$15.9 \times 4 =$

4 Divide. Use words, pictures, or equations to show your work

$88 \div 2 =$

WEDNESDAY Fractions

1 Add. Show your work.

$$3\frac{5}{6} + 5\frac{2}{3} =$$

2 Divide.

$$\frac{1}{7} \div 5 =$$

3 Plot $\frac{1}{8}$ on the number line. Is it closest to 0, $\frac{1}{2}$, or 1?

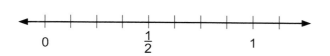

4 Dominic made 8 servings of chocolate pudding. If he made $\frac{1}{3}$ as many servings of rice pudding as chocolate pudding, how many servings of rice pudding did he have? Simplify your answer. Use a model and show your thinking.

THURSDAY Geometry

1 Sort the shapes below into the chart.

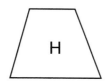

Property	Shapes
Quadrilaterals	
Parallelograms	

2 Which shapes do not belong in either group?

_____

FRIDAY Measurement and Data

1 4 years = _____ days

2 If the perimeter of a hexagon is 60 cm, what is the length of each side?

3 a) What is the volume of this shape in cubic units?

 _____

b) What would be the total volume of 10 shapes like this one?

_____

4 A box measures 5 inches by 3 inches by 2 inches.

a) Find the area of the base.

_____ × _____ = _____

b) Multiply the area of the base by the height.

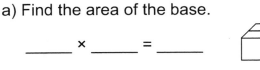

BRAIN STRETCH

Alex has $\frac{2}{4}$ quarts of chocolate milk left in a jug. In another jug, he has $2\frac{1}{2}$ pints of chocolate milk. Can Alex pour the milk from both jugs into a third jug that has a capacity of 2 gallons? Justify your answer.

MONDAY — Operations and Algebraic Thinking

1 a) Write the first five terms of each pattern.

Start at 1 and add 2 each time.

_____, _____, _____, _____, _____

Start at 1 and add 3 each time.

_____, _____, _____, _____, _____

b) Compare the two patterns. What do you notice?

_____

_____

2 7 + (7 × 30 − 19) + 28 =

3 Write an algebraic expression for this description:

b increased by 11

TUESDAY — Operations in Base Ten

1 _____ hundreds = 30,000 ones

2 Write the number in standard form.

7,000,000 + 300,000 + 70,000

+ 3,000 + 80 + 2 =

3 Compare the decimals using <, >, or =.

0.999 ☐ 0.009

4 Multiply. Use words, pictures, or equations to show your work.

6.3 × 5 =

WEDNESDAY — Fractions

1 Add.

$$\frac{6}{7} + \frac{1}{14} =$$

2 Write the mixed number as an improper fraction.

$$2\frac{4}{5}$$

3 Use a model to find $\frac{4}{5} \times 5$.

4 Four friends shared 7 peanut butter sandwiches equally. How many sandwiches did each friend eat? Express your answer as a fraction. Show your work using pictures, words, and numbers.

THURSDAY — Geometry

1 Plot the points on the coordinate plane.

A (1, 2)   E (1, 5)
B (6, 1)   F (1, 3)
C (1, 8)   G (1, 9)
D (6, 4)   H (6, 7)

2 Join points A, C, E, F, and G. What do you see?

3 Join points B, D, and H. What do you see?

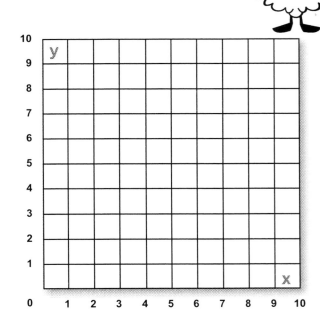

Week 13

1 9 yards = _____ feet

2 How many dimes are in $20?

3 Corey was in a track and field competition. He completed the 2,000-meter walk in 9.5 minutes. How many meters did he walk per minute?

4 a) Each cube has sides 1 inch. What is the total volume of the shape?

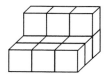

_____

b) What would be the total volume of 3 shapes like this one?

_____

BRAIN STRETCH

Jennifer saved $30. Last week she spent half of her savings on new clothes and one eighth of her savings on music downloads. What fraction of her savings does she have left? How much money does she have left?

MONDAY — Operations and Algebraic Thinking

a) Copy the patterns from Monday Question 1 on page 37.

_____, _____, _____, _____, _____

_____, _____, _____, _____, _____

b) Use the terms from each pattern to write ordered pairs.

1st term (_____ , _____)

2nd term (_____ , _____)

3rd term (_____ , _____)

4th term (_____ , _____)

5th term (_____ , _____)

c) Plot the ordered pairs on the coordinate plane and join the points. What do you see?

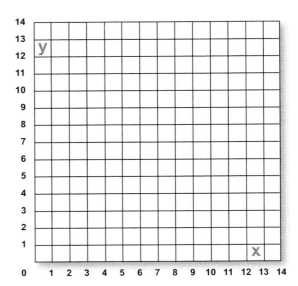

TUESDAY — Operations in Base Ten

1 $9{,}457 \div 10^2 =$

2 What is the place value of the 7 in 0.67?

3 Complete the sentence using <, >, or =.

9.6 ☐ 0.3 + 9.3

4 Round 4.317 to the nearest hundredth.

5 Write the number in standard form.

$5 \times 1{,}000 + 8 \times 100 + 2 \times 10 + 9 \times 1 + 4 \times (\frac{1}{10}) + 2 \times (\frac{1}{100}) + 9 \times (\frac{1}{1{,}000})$

=

WEDNESDAY Fractions

1 Add, then simplify your answer.

$$\frac{1}{2} + \frac{6}{7} =$$

2 Write the mixed number as an improper fraction.

$$5\frac{3}{4}$$

3 Compare the quantities using <, >, or =.

$$5 \times \frac{1}{5} \boxed{} \frac{1}{5}$$

4 Anna plays a computer game for $\frac{4}{5}$ of an hour each evening. How many hours does Anna spend playing computer games in a week? In 8 weeks? Use a model or an equation to show your work.

THURSDAY Geometry

1 Choose all the words that describe this shape.

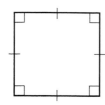

    A. rectangle

    B. quadrilateral

    C. square

    D. parallelogram

2 Which of these quadrilaterals are trapezoids?

A.   B.   C.

3 Draw an irregular polygon with 5 sides.

4 Draw and name a triangle that has no right angles.

Measurement and Data

Students at Elmwood Public School performed a play on four days.
This graph shows the number of tickets sold for each performance.

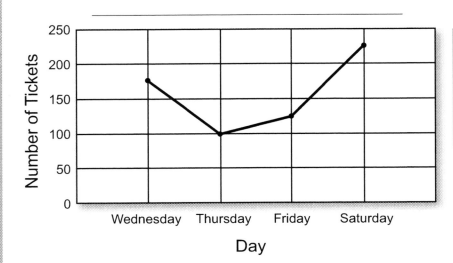

Day	Number of Tickets
Wednesday	
Thursday	
Friday	
Saturday	

1 Use the graph to complete the data table.

2 For which day were the most tickets sold? _____

3 What was the range of the number of tickets sold? _____

4 What was the increase in the number
of tickets sold from Friday to Saturday? _____

5 Add a title to the graph.

BRAIN STRETCH

Luke wants to bake some ginger cookies, but he only wants to make half of the recipe.
The recipe calls for $1\frac{1}{3}$ cups of sugar. How much sugar should Luke use?

MONDAY — Operations and Algebraic Thinking

1 [(3 + 4 + 13) × 2] ÷ (3 + 5) =

2 Write an algebraic expression for this description:

multiply the sum of 5 and 7 by the sum of 2 and 3

3 Classify the numbers as prime (P) or composite (C).

A. 22 _____

B. 11 _____

4 Find the prime factorization of 7.

TUESDAY — Operations in Base Ten

1 Round 535,871.76 to the nearest thousand.

2 What is the place value of the 3 in 14,519.73?

3 Divide. Use words, pictures, or equations to show your work

90 ÷ 3 =

4
$$\begin{array}{r} 3.8 \\ \times\ 4.9 \\ \hline \end{array}$$

WEDNESDAY — Fractions

1 Write the improper fraction as a mixed number.

$$\frac{34}{16}$$

2 Subtract, then simplify your answer.

$$\frac{6}{8} - \frac{2}{4} =$$

3 Use a model to find $3 \times \dfrac{1}{5}$.

4 Sarah bought $5\frac{1}{2}$ L of lemonade for a party. After the party, $2\frac{1}{4}$ L of lemonade were left. How much lemonade did the party guests drink?

THURSDAY — Geometry

1 a) Plot the points on the coordinate plane.

A (2, 9) B (8, 9) C (8, 4) D (2, 4)

b) Join the points.
What shape do you see?

_____

2 a) Cover the coordinate plane. Circle the coordinate pairs that you predict will be inside the shape.

(3, 5)   (5, 8)   (10, 2)
(7, 1)   (6, 6)   (9, 9)

b) Plot the points on the coordinate plane to check your predictions.

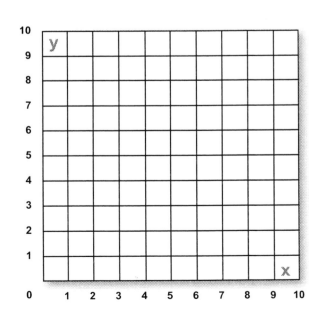

Chalkboard Publishing © 2012
Week 15

1 _____ cm = 13,000 mm

2 Patricia paid $38.52 for 9.5 gallons of gas. What was the price of each gallon of gas?

3 What are the perimeter and area of a square garden with sides 9 meters long?

4 a) Find the length, width, and height of the rectangular prism. Each cube has sides 1 cm.

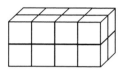

b) Use your answers in part a) to find the volume of the shape.

BRAIN STRETCH

A car left Sprucetown and traveled for 4 hours and 32 minutes to Pine City. Then the car traveled for 2 hours and 35 minutes, and arrived in Oakville at 3:30 p.m. What time did the car leave Sprucetown? Show your work.

MONDAY — Operations and Algebraic Thinking

1 a) Write the first five terms of each pattern.

Start at 12 and subtract 1 each time.

_____, _____, _____, _____, _____

Start at 12 and subtract 3 each time.

_____, _____, _____, _____, _____

b) Compare the two patterns. What do you notice?

_____

_____

2 63 ÷ 9 × (5 − 3) =

3 Write an algebraic expression for this description:

divide 16 by 8, then add 43

TUESDAY — Operations in Base Ten

1 4 ten thousands = _____ tens

2 Which digit is in the hundred thousands place?

1,724,683   _____

3 Complete the sentence using <, >, or =.

0.62 ☐ 8.9 − 6.0

4 Write the number in expanded form.

675.398 =

WEDNESDAY Fractions

1 Add.

$$14 + \frac{3}{4} =$$

2 Compare the quantities using <, >, or =.

$$3\frac{1}{3} \quad \boxed{} \quad \frac{12}{3}$$

3 Use a model to divide $\frac{1}{8} \div 3$.

4 A recipe calls for half a tablespoon of sugar per serving. How many tablespoons of sugar do you need to make 8 servings? Use a model to show your work.

THURSDAY Geometry

1 Mark all the pairs of parallel sides in these shapes.

A   B   C   D   E   F

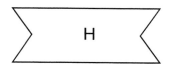

G

H

2 Sort the shapes into the chart.

Property	Shapes
no parallel sides	
1 pair of parallel sides	
2 pairs of parallel sides	

Ray sorted the nails in his toolbox by length.
This line plot shows all his nails by length.

Nails Sorted by Length (in inches)

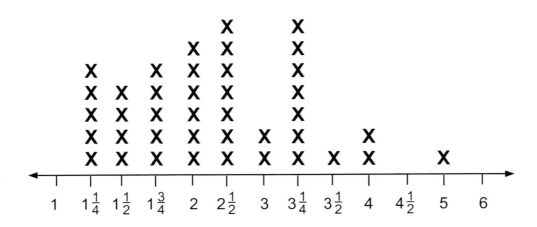

1 Ray found three 4-inch nails and three $4\frac{1}{2}$-inch nails in the garage.
He found a 2-inch nail in the bottom of a drawer. Add this data to the line plot.

2 How many nails does Ray have altogether? _____

3 Ray needs five 2-inch nails to hang some pictures. Does he have enough? _____

4 Ray needs six $3\frac{1}{4}$-inch nails and six 5-inch nails
to put together a bookcase. Does he have enough? _____

If not, what does he need more of? _____

BRAIN STRETCH

Natalia needs a gift box with a volume of 240 cubic inches and a height of 8 inches.
What are three possible combinations for the width and length of such a box?
Use whole numbers (no fractions or decimals).

MONDAY — Operations and Algebraic Thinking

a) Copy the patterns from Monday Question 1 on page 46.

_____, _____, _____, _____, _____

_____, _____, _____, _____, _____

b) Use the terms from each pattern to write ordered pairs.

1st term (_____ , _____)

2nd term (_____ , _____)

3rd term (_____ , _____)

4th term (_____ , _____)

5th term (_____ , _____)

c) Plot the ordered pairs on the coordinate plane and join the points. What do you see?

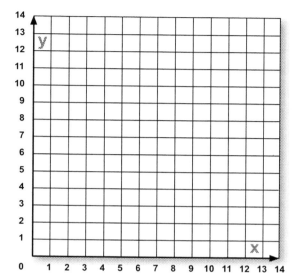

TUESDAY — Operations in Base Ten

1 6 ten thousands = _____ ones

2 a) $216 \div 10^4 =$

b) How does the exponent 4 relate to the number of zeroes in the quotient?

3 Complete the sentence using <, >, or =.

0.8 ☐ 7.9 − 7.0

4 Make the greatest possible number with 2 decimal places using these digits.

7 9 4 8 7 6 2

WEDNESDAY Fractions

1 Write the improper fraction as a mixed number.

$$\frac{13}{4}$$

2 Subtract. Show your work.

$$6\frac{1}{2} - 3\frac{4}{6} =$$

3 Complete the fraction to make the equation true.

$$3 \times \frac{\quad}{3} = 1$$

4 On Thursday, $\frac{1}{6}$ of the students in the class were absent. If there are 36 students in total, how many students were absent? Use a model to show your work.

THURSDAY Geometry

1 Write the coordinates of the vertices of the rectangle.

A _____ B _____ C _____ D _____

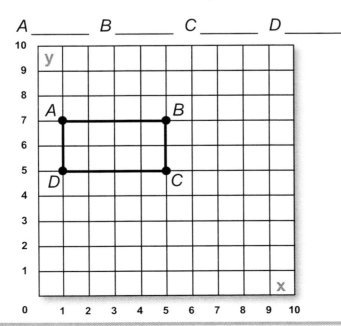

2 What is the length of the rectangle? _____ units

What is the width of the rectangle? _____ units

3 Draw a new rectangle that is twice as wide and twice as long as rectangle *ABCD*.

Label the vertices of your rectangle *M, N, O, P*.

Write the coordinates of the vertices below.

M _____ N _____

O _____ P _____

1 9 kg = _____ g

2 How many nickels are in $10?

3 George has a paper airplane that can fly 3.6 m. His friend Paul has one that can go 246 cm. Whose paper airplane can fly farther? How much farther can that person's airplane fly in centimetres?

4 Each cube has sides 1 meter. What is the total volume of the shape?

_____

BRAIN STRETCH

Ann Marie bought 2 bags of popcorn for $2.58 each and a drink for $1.20. She paid with $7.00. What are two combinations of coins that Ann Marie could receive as change? What is the fewest number of coins she could receive?

MONDAY Operations and Algebraic Thinking

1 Write an algebraic expression for this description:

take 40 away from 50, then multiply by 8

2 Evaluate the expression for $k = 9$.

$$20 - k$$

3 How many times larger or smaller? Compare the expressions without calculating.

$2 \times (4 + 15)$ is _____ times

_____ than $4 \times (4 + 15)$

4 Find the prime factorization of 48.

TUESDAY Operations in Base Ten

1 Round 234,983.3 to the nearest hundred.

2 a) $3.5 \times 10^3 =$
   b) How does the exponent 3 relate to the number of zeroes in the product?

3 Round 13.259 to the nearest tenth.

4 $985.2 \div 100 =$

5 Write the number in standard form.

$$4 \times 100 + 2 \times 10 + 6 \times \left(\frac{1}{10}\right) + 7 \times \left(\frac{1}{100}\right) + 9 \times \left(\frac{1}{1,000}\right)$$

$=$

WEDNESDAY Fractions

1 Add, then simplify your answer.

$$\frac{4}{5} + \frac{3}{15} =$$

2 Is $\frac{5}{8}$ closest to 0, $\frac{1}{2}$, or 1?

3 Divide.

$$5 \div \frac{1}{4} =$$

4 Curtis's aunt is $3\frac{1}{2}$ times older than he is. If Curtis is 9, how old is his aunt?

THURSDAY Geometry

1 Choose all the words that describe this shape.

    A. rectangle

    B. quadrilateral

    C. rhombus

    D. trapezoid

2 Classify the triangle by the lengths of its sides.

    7 ft.,   7 ft.,   7 ft.

    A. isosceles

    B. scalene

    C. equilateral

3 Name and draw 2 special quadrilaterals.

4 Are these shapes congruent or similar?

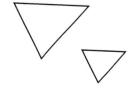

1 5 pints = _____ cups

2 Daryl drank 6 bottles of water each day for two weeks. Each bottle held 500 milliliters of water. How many liters of water did Daryl drink per week?

3 Calculate the perimeter and area of the polygon.

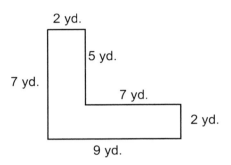

4 The volume of each box is 16 cubic inches. What is the volume of the whole 3D shape?

_____

BRAIN STRETCH

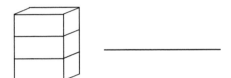

Pedram paid $30 for a new video game. He had $50 in savings. He says he spent $\frac{1}{4}$ of his savings on the video game. Is Pedram correct? Justify your answer. If Pedram is not correct, find the fraction of his savings he actually spent on the game.

MONDAY — Operations and Algebraic Thinking

1 a) Write the first five terms of each pattern.

Start at 4 and add 2 each time.

_____, _____, _____, _____, _____

Start at 0 and add 2 each time.

_____, _____, _____, _____, _____

b) Compare the two patterns. What do you notice?

_____

_____

2 Write an algebraic expression for this description:

add 3 and 16, then multiply by 5

3 How many times larger or smaller? Compare the expressions without calculating.

1,000 × (934 − 108) is _____ times

_____ (934 − 108) × 10

TUESDAY — Operations in Base Ten

1 7 hundred thousands

= _____ ten thousands

2 What is the value of the underlined digit?

43,428,263   _____

3 Estimate the difference by rounding each number to the nearest whole number and then subtracting.

9.27 − 6.946

4
$$\begin{array}{r} 4.4 \\ \times\ 1.6 \\ \hline \end{array}$$

WEDNESDAY — Fractions

1 Write the improper fraction as a mixed number.

$$\frac{17}{12}$$

2 Subtract.

$$15\frac{1}{2} - 4 =$$

3 Each square in the grid has sides that measure $\frac{1}{3}$ m. What is the area of the rectangle in meters?

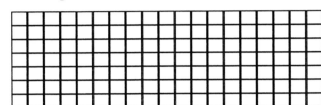

4 Megan went to the store with $5. She spent $\frac{1}{5}$ of her money on a fruit smoothie. How much did Megan spend on the smoothie?

THURSDAY — Geometry

What are the coordinates of these places?

 Bakery _____

Bank _____

Flower shop _____

Gas station _____

Pet store _____

Restaurant _____

In the City

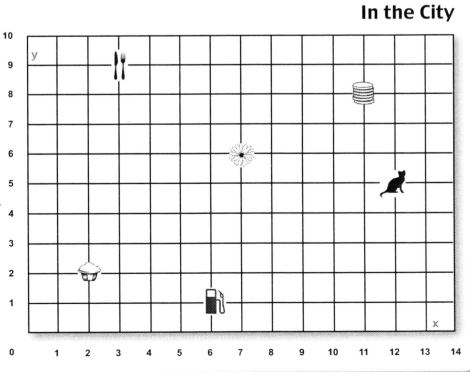

Mr. Brant's students sold cookies to raise money for a field trip.

Number of Boxes of Cookies Sold by Each Student

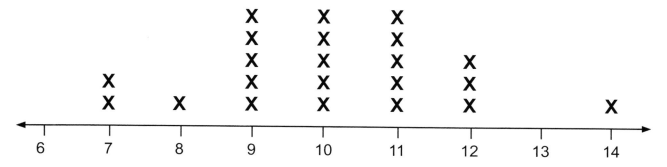

1 How many boxes of cookies did the students sell altogether? _____

2 If the class receives $2 for each box of cookies sold, how much money did the students raise? _____

3 Students who sell more than 10 boxes receive a prize from the cookie company. How many students in Mr. Brant's class received a prize? _____

BRAIN STRETCH

A garden is 8 feet longer than three times its width.
Let *w* represent the width of the garden and let *l* represent the length of the garden.

a) Write an expression for the length of the garden.
b) What an expression for the perimeter of the garden.

MONDAY · Operations and Algebraic Thinking

a) Copy the patterns from Monday Question 1 on page 55.

_____ , _____ , _____ , _____ , _____

_____ , _____ , _____ , _____ , _____

b) Use the terms from each pattern to write ordered pairs.

1ˢᵗ term (_____ , _____)

2ⁿᵈ term (_____ , _____)

3ʳᵈ term (_____ , _____)

4ᵗʰ term (_____ , _____)

5ᵗʰ term (_____ , _____)

c) Plot the ordered pairs on the coordinate plane and join the points. What do you see?

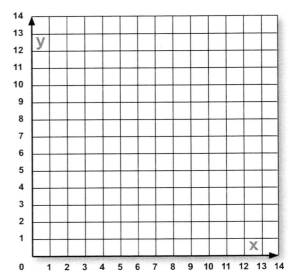

TUESDAY · Operations in Base Ten

1 Round 56.98 to the nearest one.

2 Estimate the sum by rounding each number to the nearest whole number and then adding.

6.23 + 1.754

3
```
  74.53
- 13.06
```

4
```
   80
 × 84
```

WEDNESDAY — Fractions

1 Write the mixed number as an improper fraction.

$8\dfrac{10}{11}$

2 Add.

$10 + 2\dfrac{3}{7} =$

3 Is $\dfrac{1}{5}$ closest to 0, $\dfrac{1}{2}$, or 1?

4 A bag of flour that weighs $\dfrac{1}{2}$ lb is used to make 4 batches of bread. How much flour does each batch of bread use?

THURSDAY — Geometry

1 Choose all the words that describe this shape.

A. parallelogram
B. 4 sides
C. quadrilateral
D. rhombus

2 Name and draw a shape that has only one pair of parallel sides.

3 Classify the triangle.

A. isosceles
B. scalene
C. right
D. acute
E. obtuse

4 How many lines of symmetry are there?

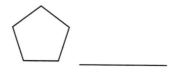

_____

1 6.7 liters = _____ milliliters

2 a) Find the length, width, and height of the rectangular prism. Each cube has sides 1 inch.

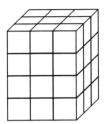

b) Use your answers in part a) to find the volume of the shape.

3 Darius built a chest for his sports equipment. The chest is 4 feet long, 3 feet wide, and 2 feet high. He painted the outside of the chest, but not the bottom. What area did Darius paint?

4 What is the volume of a bathtub that is 6 feet long, 2 feet wide, and 1 foot deep?

BRAIN STRETCH

Brendan made 7 pints of lemonade for a class party.
Carla made 8 more pints of lemonade for the class party.

a) How many cups of lemonade did Brendan and Carla make altogether?
b) If there are 20 students in the class, is there enough lemonade for each student to have 3 cups? Justify your answer.

1 $(9 + 30) \div (5 - 2) =$

2 Write an algebraic expression for this description:

subtract 5 from 11, then multiply by 2

3 How many times larger or smaller? Compare the expressions without calculating.

$(154 \times 3) \times 10$ is _____ times

_____ than $(154 \times 3) \times 2$

4 List the first 5 multiples of the number 7.

_____ _____ _____ _____ _____

1 2 thousands = _____ tens

2 Round 64.359 to the nearest hundredth.

3 Make the least possible number with 3 decimal places using these digits.

2 5 9 4 8 1

4 $3\overline{)408}$

WEDNESDAY Fractions

1 Write the mixed number as an improper fraction.

$3\dfrac{3}{7}$

2 Add.

$\dfrac{17}{4} + \dfrac{5}{2} =$

3 Each square in the grid has sides that measure $\frac{1}{5}$ yd. What is the area of the rectangle in yards?

4 Adam ate $\frac{3}{8}$ of his mom's apple pie. If it had 12 pieces, how many pieces did Adam eat?

THURSDAY Geometry

1 Plot these points on the coordinate plane.

A (6, 9) B (6, 1)

2 Join the points to create line AB.

3 Draw a line that is perpendicular to line AB. Write the coordinate pairs for two points on the perpendicular line:

_____   _____

4 Draw a line that is parallel to line AB. Write the coordinate pairs for two points on the parallel line:

_____   _____

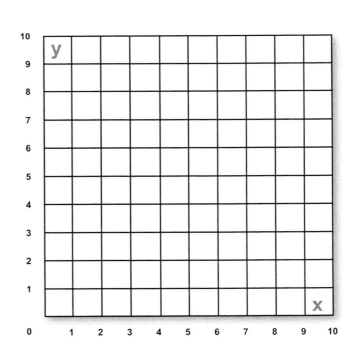

1 a) 1 hour = _____ seconds

b) 8 hours = _____ minutes

c) _____ hours = 1,020 minutes

2 The volume of each rectangular prism is 2 cubic meters. What is the volume of the whole 3D shape?

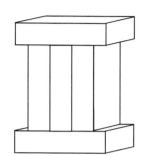

3 Mrs. Simpson needs to buy sand for the school sandbox. She buys 336 ft.³ of sand for a sandbox that is 18 feet wide and 1 foot deep. How long is the sandbox?

BRAIN STRETCH

a) It costs $0.04 per minute to make a phone call. Adam was on the phone for 15 minutes. How much did his phone call cost? Write an equation to find the cost of the phone call.

b) Adam's phone card has a value of $5. How many 15-minute phone calls can Adam make before the card runs out of money?

MONDAY — Operations and Algebraic Thinking

1 a) Write the first five terms of each pattern.

Start at 1 and add 2 each time.

_____, _____, _____, _____, _____

Start at 3 and add 1 each time.

_____, _____, _____, _____, _____

b) Compare the two patterns. What do you notice?

_____

_____

2 Write an algebraic expression for this description:

a plus 50

3 Classify the numbers as prime (P) or composite (C).

A. 56 _____

B. 39 _____

TUESDAY — Operations in Base Ten

1 7 hundreds = _____ ones

2 3,400 ÷ 100 =

3 Write the value of the underlined digit.

406,7̲34.35   _____

4
$$\begin{array}{r} 29.08 \\ + 65.75 \\ \hline \end{array}$$

WEDNESDAY — Fractions

1 Subtract. Show your work.

$$4\frac{6}{8} - 1\frac{2}{3} =$$

2 Divide.

$$\frac{1}{12} \div 6 =$$

3 Use a model to find $\frac{3}{7} \times 4$.

4 Rania picked $1\frac{2}{3}$ pints of strawberries on Monday, $1\frac{1}{2}$ pints on Tuesday, and $1\frac{3}{4}$ pints on Wednesday. How many pints of strawberries did Rania pick altogether

THURSDAY — Geometry

What are the coordinates of the following animals?

Bison _____

Crocodiles _____

Elephants _____

Gorillas _____

Hippopotamuses _____

Leopards _____

At the Zoo

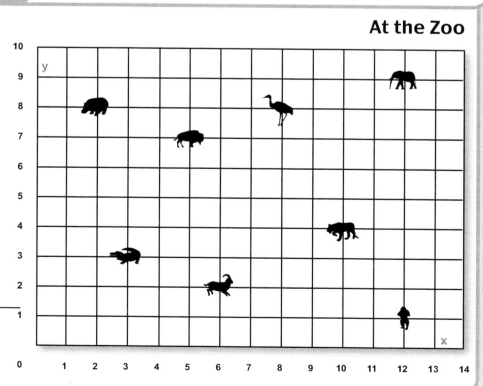

1 Construct a horizontal bar graph using the information from the table.

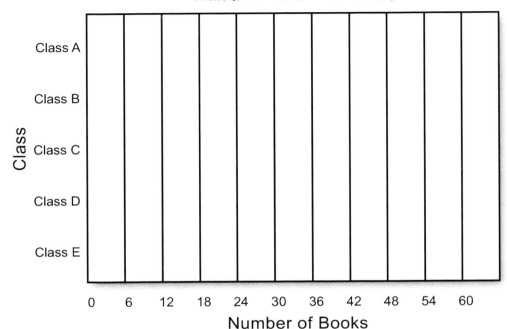

Number of Books Read

Class	Number of Books
Class A	50
Class B	43
Class C	63
Class D	38
Class E	57

2 Which class read the most books? _____

3 If each student in Class E read 3 books, how many students are in class E? _____

4 How many books did the classes read altogether? _____

5 Which classes read more than 45 books? _____

BRAIN STRETCH

How many squares can you find in the picture?

a) Copy the patterns from Monday Question 1 on page 64.

_____, _____, _____, _____, _____

_____, _____, _____, _____, _____

b) Use the terms from each pattern to write ordered pairs.

1st term (_____ , _____)

2nd term (_____ , _____)

3rd term (_____ , _____)

4th term (_____ , _____)

5th term (_____ , _____)

c) Plot the ordered pairs on the coordinate plane and join the points. What do you see?

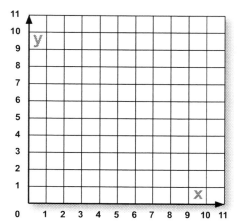

1 To which place value was the number rounded?

$134{,}867 \rightarrow 135{,}000$

_____

2 Write the number in expanded form.

$778{,}145.29 =$

3
```
   51.74
+ 32.38
```

4
```
   230
×   14
```

WEDNESDAY — Fractions

1 Write the improper fraction as a mixed number.

$$\frac{9}{5}$$

2 Complete the fraction to make the equation true.

$$4 \times \frac{3}{\underline{\hspace{1cm}}} > 4$$

3 Find the area of the square.

$\frac{2}{3}$ ft.

4 Sue has a collection of 42 stuffed animals. She decided to donate $\frac{2}{3}$ of them to a children's charity. How many stuffed animals does Sue have left?

THURSDAY — Geometry

1 Choose all the words that describe this shape.

A. polygon
B. hexagon
C. pentagon
D. octagon

2 Draw and name 3 types of triangles.

3 Classify the triangle by the lengths of its sides.

5 ft.,   20 ft.,   20 ft.

A. isosceles
B. scalene
C. equilateral

4 Draw a line of symmetry.

T

1 33.52 L = _____ mL

2 A hot tub is 7 feet long and 6 feet wide. It holds 126 ft.³ of water. How deep is the tub?

3 What are the perimeter and area of a tabletop with width 60 inches and length 120 inches? Will a tablecloth with an area of 600 in.² fit on the tabletop? Justify your answer

4 Each week Samantha practices gymnastics for 445 minutes. How many hours and minutes does she practice gymnastics?

BRAIN STRETCH

Lexi is making a smoothie with yogurt, fruit, and orange juice. She used $1\frac{3}{4}$ cups more fruit than yogurt and $1\frac{1}{2}$ cups more juice than fruit. If she used half a cup of yogurt, how many cups of smoothie did Lexi make altogether? If she shared the smoothie equally with two other friends, what was the size of each serving? Express your answers as fractions. Show your work using pictures, words, and numbers.

1 [11 − (4 + 2)] × 3 =

2 Write an algebraic expression for this description:

m less than 97

3 Solve using the inverse operation.

44 − x = 20

x = _____

4 Find the prime factorization of 34.

1 1 million = _____ thousands

2 Write thirty-two hundredths as a decimal.

3 Divide. Use words, pictures, or equations to show your work.

63.3 ÷ 3 =

4
$$\begin{array}{r} 9.2 \\ \times\ 3.3 \\ \hline \end{array}$$

WEDNESDAY　Fractions

1 Add.

$$\frac{13}{2} + 1\frac{1}{3} =$$

2 Compare the fractions using <, >, or =.

$$\frac{13}{4} \boxed{} \frac{13}{2}$$

3 Divide.

$$18 \div \frac{1}{2} =$$

4 Ahmed used one quarter of a gallon of milk to make 2 batches of ice cream. How many gallons of milk did Ahmed put in each batch of ice cream

THURSDAY　Geometry

1 Mark all the right angles in these shapes.

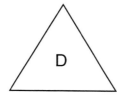

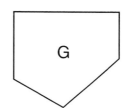

A　B　C　D　E　F　G　H

2 Sort the shapes into the chart.

Property	Shapes
0 right angles	
1 right angle	
2 right angles	
4 right angles	

Each family on Chester Street has a rain barrel. This list shows the amount of water in each rain barrel at the end of the week:

Water Collected (in liters)

8    $8\frac{1}{2}$ 10   11   6   8 $9\frac{1}{2}$ $8\frac{1}{2}$ 7

1 Create a line plot of the data.

6   12

2 What is the total amount of water collected in the rain barrels? _____

3 How many 4-liter watering cans can be filled with the water? _____

How much water is left over? _____

4 If the families want to share the water equally,
how many cans of water will each family get? _____

BRAIN STRETCH

You have a rectangular backyard. It is 5 yards long and 30 yards wide.
It costs $1,250 to put a wooden fence around the backyard.
A bag of grass seed costs $5.99 and covers 50 square yards.

a) How many bags of grass seed do you need to seed the backyard?
b) How much will it cost to seed the backyard and put up the fence?

MONDAY — Operations and Algebraic Thinking

1 Write an algebraic expression for this description:

the quotient of seventy and five more than two

2 Solve using the inverse operation.

$21 + x = 78$

$x = $ _____

3 How many times larger or smaller? Compare the expressions without calculating.

$46{,}024 \times 12$ is _____ times

_____ than $46{,}024 \times 3$

4 Find the common factors of 8 and 18.

TUESDAY — Operations in Base Ten

1 $22 \times 10^4 =$

2 Write 872.549 in words.

3 What is the place value of the 1 in 96.741?

4
$$\begin{array}{r} 43 \\ \times\ 94 \\ \hline \end{array}$$

5 Write the number in standard form.

$7 \times 100 + 5 \times 10 + 2 \times 1 + 8 \times (\frac{1}{10}) + 6 \times (\frac{1}{100}) + 2 \times (\frac{1}{1{,}000})$

$=$

1 Write the mixed number as an improper fraction.

$$9\frac{1}{8}$$

2 Complete the fraction to make the equation true.

$$7 \times \frac{1}{\underline{}} = 1$$

3 Sarah went on a hike that was $2\frac{1}{3}$ miles long. The next day Sarah went on a hike that was four times as long. How far did Sarah hike altogether?

4 Henry spent 3 hours at the museum. He spent $\frac{1}{5}$ of the time at the dinosaur exhibit and the rest of the time at the moon exhibit. How much time did Henry spend at each exhibit?

THURSDAY — Geometry

a) What are the coordinates of the events?

At the Summer Olympic Games

Archery _____

Cycling _____

Equestrian Jumping _____

Gymnastics _____

Soccer _____

Weightlifting _____

b) Suzanne started at the soccer event and walked 4 units to the right and 2 units up. What event did she go to?

_____

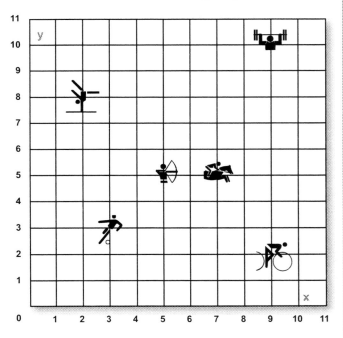

1 20 minutes = _____ seconds

2 Samantha brought 2 pounds of jelly beans to a class party. There are 24 students and one teacher in the class. Can Samantha give each person 2 ounces of jelly beans? Justify your answer.

3 What are the perimeter and area of a room 18 yards wide and 20 yards long in inches?

4 Find the volume and the surface area of the rectangular prism.

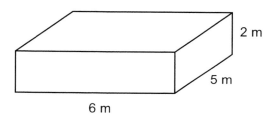

2 m

5 m

6 m

BRAIN STRETCH

Spencer's cell phone company charges him $39.99 a month for up to 400 minutes, and 20 cents per minute for every minute over 400. If n represents the number of minutes over 400, this is the expression the cell phone company uses to calculate Spencer's bill: 39.99 + 0.20n. Use the expression to calculate Spencer's bill if he talked for 425 minutes last month.

Operations and Algebraic Thinking

1 Write an algebraic expression for this description:

subtract 10 from 100, then divide by 9

2 Evaluate the expression for $x = 9$.

$11 + x$

3 How many times larger or smaller? Compare the expressions without calculating.

$75 \times 2{,}430$ is _____ times

_____ than $25 \times 2{,}430$

4 Find the prime factorization of 11.

Operations in Base Ten

1 Write the value of the underlined digit.

37,890.<u>5</u>6 _____

2 Estimate the difference by rounding each number to the nearest whole number and then subtracting.

$5.67 - 1.93$

3 Divide. Use words, pictures, or equations to show your work.

$1 \div 0.2 =$

4 $8 \overline{)896}$

WEDNESDAY Fractions

1 Write the improper fraction as a mixed number.

$$\frac{49}{11}$$

2 Find the area of the rectangle.

$\frac{5}{8}$ m

$\frac{7}{8}$ m

3 Divide.

$$4 \div \frac{1}{5} =$$

4 There are $52\frac{1}{2}$ gallons of water in your rain barrel. Your mom uses 11 gallons to water the garden, your brother uses $3\frac{1}{2}$ gallons to wash his bike, and your sister uses $8\frac{1}{2}$ gallons to give the dog a bath. How many gallons of water are left in the rain barrel?

THURSDAY Geometry

1 Choose all the words that describe this shape.

A. parallelogram
B. 4 sides
C. quadrilateral
D. trapezoid

2 Classify the triangle. Circle all the descriptions that apply.

A. acute
B. isosceles
C. obtuse
D. equilateral
E. right

3 How many lines of symmetry are there?

Π _____

4 Can you construct a triangle that is both a right triangle and an isosceles triangle? If yes, draw an example. If no, explain why this combination is not possible.

1 _____ kg = 1,200 g

2 The volume of each brick in this wall is 6 cubic feet. What is the volume of the whole wall?

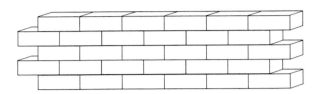

3 A rectangular prism has a length of 22 inches, a height of 8 inches, and a width of 3 inches. What is its volume?

4 Here is the design for the new art room at Mina's school.

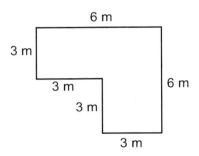

6 m

3 m

3 m

3 m

6 m

3 m

a) What is the perimeter of the new art room?
b) How much carpet will the school need to buy to cover the floor in the art room?

BRAIN STRETCH

Mathew and Christie picked 176 ounces of berries in order to bake some pies.

a) How many pounds of berries did they pick?
b) If they need $1\frac{1}{2}$ pounds of berries per pie, how many pies can they bake? Show your work.

a) Write the first five terms of each pattern.

Start at 0 and add 3 each time.

_____, _____, _____, _____, _____

Start at 0 and add 9 each time.

_____, _____, _____, _____, _____

b) Compare the two patterns. What do you notice?

_____

_____

c) Use the terms from each pattern to write ordered pairs.

1st term (_____ , _____)

2nd term (_____ , _____)

3rd term (_____ , _____)

4th term (_____ , _____)

5th term (_____ , _____)

d) Predict what you would see if you plotted the ordered pairs on a coordinate plane. Use grid paper to check your prediction.

1 $85 \times 100 =$

2 Write the value of the underlined digit.

6<u>3</u>2,789.15 _____

3
$$\begin{array}{r} 7.5 \\ \times\ 8.7 \\ \hline \end{array}$$

4 $90\overline{)4,950}$

5 Write the number in standard form.

$2 \times 100 + 7 \times 10 + 4 \times 1 + 3 \times (\frac{1}{10}) + 8 \times (\frac{1}{100}) + 6 \times (\frac{1}{1,000})$

=

WEDNESDAY — Fractions

1 Subtract.

$$12 - \frac{3}{5} =$$

2 Divide.

$$9 \div \frac{1}{6} =$$

3 Find the area of the square.

$\frac{1}{5}$ yd.

4 Samantha the electrician wired 6 rooms in a house. If each room needed $\frac{3}{4}$ of a roll of electrical wire, how many rolls did Samantha use?

THURSDAY — Geometry

1 Classify the triangle by the lengths of its sides.

9 ft.,   12 ft.,   14 ft.

A. isosceles
B. scalene
C. equilateral

2 Can an equilateral triangle be obtuse? Show your thinking.

3 How many lines of symmetry are there?

B _____

4 True or false: Every parallelogram is a quadrilateral, but not every quadrilateral is a parallelogram. Use examples to support your answer.

Week 27

The students at Orchard Park Public School held a canned food drive.
This table shows the number of cans they collected each day over three weeks.

	Week 1	Week 2	Week 3
Monday	11	28	8
Tuesday	23	44	13
Wednesday	30	40	13
Thursday	18	33	28
Friday	11	15	8

1 What was the total number of cans collected each week?

Week 1 _____   Week 2 _____   Week 3 _____

2 What was the mean number of cans collected per day each week?

Week 1 _____   Week 2 _____   Week 3 _____

3 In which week did students collect the most cans? _____
Describe one reason why students might have collected more cans during this week than the others.

_____

_____

4 Compare the number of cans collected on Mondays and Fridays to the number collected on other days of the week. What pattern do you see? Describe one possible reason for the pattern.

_____

_____

BRAIN STRETCH

A jar holds 500 mL of jam. A carton holds 6 jars.
How many jars and cartons are needed to package 9.5 L of jam?

MONDAY — Operations and Algebraic Thinking

1 Classify the numbers as prime (P) or composite (C).

A. 3 _____

B. 60 _____

2 Write an algebraic expression for this description:

y fewer than 120

3 Compare the expressions without calculating. Use <, >, or =.

$3 + (4 \times 6)$ ⬚ $3 + (4 \times 7)$

4 Sharon took 96 flowers and split them into bunches to decorate for a party. She put 12 flowers in each bunch. Write an equation to show how many bunches of flowers Sharon made.

TUESDAY — Operations in Base Ten

1 Use $7 \times 3 = 21$ to find $63 \div 7$.

2 How does the value of the 2 in the number 345.42 compare to the 2 in the number 743.25?

3 Multiply. Use words, pictures, or equations to show your work.

$30.25 \times 10^3 =$

4
$$\begin{array}{r} 121 \\ \times 4 \\ \hline \end{array}$$

Chalkboard Publishing © 2012

1 Add. Show your work.

$$2\frac{3}{4} + 3\frac{2}{8} =$$

2 Find the product. Simplify your answer.

$$3 \times \frac{2}{10}$$

3 On Friday, Louis and his family drove 5 miles. They drove $4\frac{1}{2}$ times as far on Saturday as on Friday. How many miles did Louis's family drive on Saturday?

4 How many $\frac{2}{3}$-cup servings are in 5 cups of hot chocolate?

THURSDAY Geometry

1 Draw a rectangle with length 10 units and width 4 units. Label the vertices.

2 Draw two lines of symmetry through your rectangle.

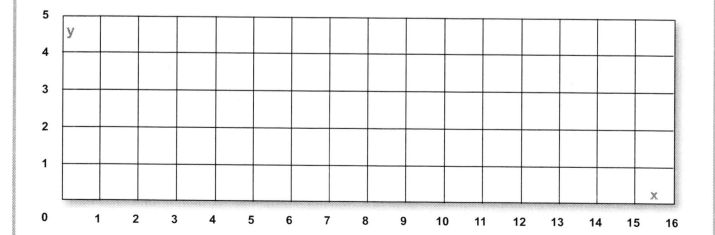

1 The volume of box A is half the volume of box B. If the volume of box A is 10 cubic feet, what is the total volume of both boxes?

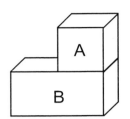

2 Dennis wants to mail birthday presents to his twin cousins. He bought two books that each weigh 95 grams, two toys that each weigh 275 grams and two cards that each weigh 25 grams. He packs everything into a box that weighs 150 grams.

a) What is the total weight of Dennis's package?

b) If the post office charges an extra $2 to mail a package over 1 kg, will Dennis have to pay the extra $2?

BRAIN STRETCH

It's pizza day at George Washington Public School! This is the number of slices of pizza the principal needs to order for each grade:

Grade 1: 40 slices
Grade 2: 49 slices
Grade 3: 63 slices
Grade 4: 75 slices
Grade 5: 65 slices

Instead of ordering slices, the principal orders whole pizzas. If each pizza is divided into 8 slices, how many pizzas does the principal need to order?

MONDAY — Operations and Algebraic Thinking

1 Find the common factors of 9 and 42.

2 Compare the expressions without calculating. Use <, >, or =.

$$(6 \times 14) - 14 \; \boxed{\phantom{<}} \; (4 \times 14) + 14$$

3 Evaluate the expression for $m = 32$.

$$m \div 4 - 6$$

4 Colin bought some stamps. He gave 22 stamps to his mom and had 8 stamps left. Write an expression for the number of stamps Colin bought.

TUESDAY — Operations in Base Ten

1 Write five hundred thirty nine thousandths as a decimal.

2 $63 \times 10^6 =$

3
$$\begin{array}{r} 86.50 \\ -\ 43.03 \\ \hline \end{array}$$

4
$$\begin{array}{r} 782 \\ \times\ \ \ 4 \\ \hline \end{array}$$

WEDNESDAY Fractions

1 Is $\frac{5}{6} \times 8$ less than or greater than 8? How do you know without multiplying?

2 Use a model to find $\frac{3}{4} \times 3$.

3 Compare the fractions using <, >, or =.

$$\frac{15}{3} \boxed{} \frac{25}{5}$$

4 Brad squeezed $\frac{5}{8}$ of a quart of orange juice for him and his two sisters to share. How much juice will each person get if the juice is shared equally?

THURSDAY Geometry

1 How is a square like a rhombus and a rectangle? How is it different?

2 Can a triangle have more than one obtuse angle? Why or why not?

3 Draw the lines of symmetry.

4 Are these shapes congruent or similar?

Measurement and Data

1 5.9 kilograms = _____ grams

2 Sarah started her homework at 7:18 p.m. and finished at 8:47 p.m. How long did Sarah spend doing her homework?

3 Farah has a special box for her treasures. The width of the box is 9 inches. The length is four times the width and the height is half the width. What is the volume of Farah's special box?

4 Find the volume and the surface area of the rectangular prism.

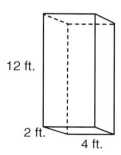

12 ft.

2 ft.

4 ft.

BRAIN STRETCH

The fifth-grade students were asked to vote for their favorite school subject.

• 12 students voted for math.
• 11 students voted for language arts.
• 3 students voted for art.
• Half as many students voted for science than for math.
• Three times as many students voted for social studies than for art.

How many students voted in the survey? Show your work.

Operations and Algebraic Thinking

a) Write the first five terms of each pattern.

Start at 0 and add 8 each time.

_____, _____, _____, _____, _____

Start at 0 and add 4 each time.

_____, _____, _____, _____, _____

b) Compare the two patterns. What do you notice?

_____

_____

c) Use the terms from each pattern to write ordered pairs.

1st term (_____ , _____)

2nd term (_____ , _____)

3rd term (_____ , _____)

4th term (_____ , _____)

5th term (_____ , _____)

d) Predict what you would see if you plotted the ordered pairs on a coordinate plane. Use grid paper to check your prediction.

Operations in Base Ten

1 What is the value of the underlined digit?

1,5<u>6</u>7,974 _____

2 Estimate the sum by rounding each number to the nearest whole number and then adding.

5.67 + 4.394

3
```
   487
 ×  35
```

4 $2\overline{)4.58}$

  Chalkboard Publishing © 2012

1 Is $\frac{1}{4} \times 7$ greater than or less than 7? How do you know without multiplying?

2 Each square in the grid has sides that measure $\frac{2}{10}$ m. What is the area of the rectangle in meters?

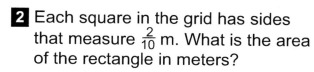

3 Divide.

$$2 \div \frac{1}{9} =$$

4 Carla's lemonade stand uses 6 bags of lemons each week. For how many days will $2\frac{1}{3}$ bags of lemons last? Justify your answer.

THURSDAY   Geometry

Look at the shapes below and sort them into groups.
Use two or more words to describe each rule for sorting.

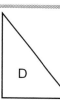

Sorting Rule	Shapes

Patricia and Tom are building shelves. This line plot shows how many wooden planks they have of each length:

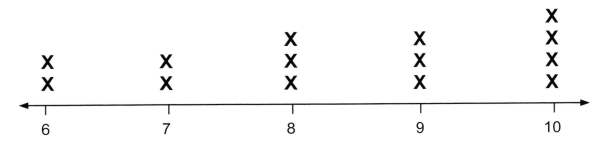

Length of Wooden Planks (in feet)

To make the shelves, they need 4 planks that are $3\frac{1}{2}$ ft. long and 6 planks that are 6 ft. long.

1 Patricia and Tom can build the bookcase using only the 6-foot planks and 10-foot planks. Explain how they can cut these planks to get the lengths they need. What pieces of wood are left over?

_____

_____

2 Describe another way that Patricia and Tom could use the planks they have to build the bookcase.

_____

_____

BRAIN STRETCH

Describe how you might measure the perimeter of a puddle.

Fraction Strips

						1

$\frac{1}{2}$	$\frac{1}{2}$

$\frac{1}{3}$	$\frac{1}{3}$	$\frac{1}{3}$

$\frac{1}{4}$	$\frac{1}{4}$	$\frac{1}{4}$	$\frac{1}{4}$

$\frac{1}{5}$	$\frac{1}{5}$	$\frac{1}{5}$	$\frac{1}{5}$	$\frac{1}{5}$

$\frac{1}{6}$	$\frac{1}{6}$	$\frac{1}{6}$	$\frac{1}{6}$	$\frac{1}{6}$	$\frac{1}{6}$

$\frac{1}{7}$	$\frac{1}{7}$	$\frac{1}{7}$	$\frac{1}{7}$	$\frac{1}{7}$	$\frac{1}{7}$	$\frac{1}{7}$

$\frac{1}{8}$	$\frac{1}{8}$	$\frac{1}{8}$	$\frac{1}{8}$	$\frac{1}{8}$	$\frac{1}{8}$	$\frac{1}{8}$	$\frac{1}{8}$

$\frac{1}{9}$	$\frac{1}{9}$	$\frac{1}{9}$	$\frac{1}{9}$	$\frac{1}{9}$	$\frac{1}{9}$	$\frac{1}{9}$	$\frac{1}{9}$	$\frac{1}{9}$

$\frac{1}{10}$	$\frac{1}{10}$	$\frac{1}{10}$	$\frac{1}{10}$	$\frac{1}{10}$	$\frac{1}{10}$	$\frac{1}{10}$	$\frac{1}{10}$	$\frac{1}{10}$	$\frac{1}{10}$

0.1	0.1	0.1	0.1	0.1	0.1	0.1	0.1	0.1	0.1

$\frac{1}{11}$	$\frac{1}{11}$	$\frac{1}{11}$	$\frac{1}{11}$	$\frac{1}{11}$	$\frac{1}{11}$	$\frac{1}{11}$	$\frac{1}{11}$	$\frac{1}{11}$	$\frac{1}{11}$	$\frac{1}{11}$

$\frac{1}{12}$	$\frac{1}{12}$	$\frac{1}{12}$	$\frac{1}{12}$	$\frac{1}{12}$	$\frac{1}{12}$	$\frac{1}{12}$	$\frac{1}{12}$	$\frac{1}{12}$	$\frac{1}{12}$	$\frac{1}{12}$	$\frac{1}{12}$

Week 1, pages 1–3

Monday 1. A 2. 47 3. $(21 + 1) \times 2$ 4. $y = 9x$

Tuesday 1. 10 2. 320 3. > 4. 1.6

Wednesday 1. Sample answer: 3/6, 4/8 2. A 3. < 4. 42/9

Thursday 1. ★ (2, 7), ✚ (2, 1), ◆ (5, 4), 🕯 (6, 8), ⚓ (7, 3), 📷 (1, 4), 🦌 (6, 2), ⬆ (9, 0), 🎿 (8, 7)

2. A circle should be drawn at (4, 9). 3. A triangle should be drawn at (0, 5).

Friday 1. 700 cm 2. B 3. They will hold the same amount. 4. 3:19 p.m.

Brain Stretch 1.56 kg

Week 2, pages 4–6

Monday 1. 2, 1, 3 2. = 3. 28 4. Double the previous number.

Tuesday 1. 600 ones 2. 8,000 3. > 4. 2.6

Wednesday 1. Sample answer: 2/16, 3/24 2. > 3. 1 7/8 4. 8 1/6 cups

Thursday 1. ○ ✐ □ ◁ ▽ ▱ ⇨ 2. ○ □ 3. ✐ ◁ ▽ ▱ ⇨

4. An open shape could be closed up, or a curved line could be straightened.

Friday 1. 1,440 oz. 2. 90 in. 3. A 4. 12

Brain Stretch 55 people

Week 3, pages 7–9

Monday 1. C 2. 66 3. $7 \times (7 - 2) + 10 = 45$ 4. Subtract 50 from the previous number.

Tuesday 1. 238,675 = 200,000 + 30,000 + 8,000 + 600 + 70 + 5 2. a) 3, 30, 300, 3,000 b) 300,000

3. 10 4. 15.1

Wednesday 1. 1 2/3 2. 11/8 3. < 4. 3 17/20

Thursday 1. C 2. A, D 3. Accept any straight-lined closed shape. 4. None

Friday 1. 8 pints 2. 445 cm 3. C

4. Sample answer: No, there is still room that is not completely taken up by the cubes.

Brain Stretch 22 slices

Week 4, pages 10–12

Monday 1. C 2. 18 3. 36, 72, 108, 144, 180 4. 17 + 20

Tuesday 1. 0.4 hundreds 2. tens of thousands 3. a) 5, 50, 500, 5,000 b) 9,000 4. 11.31

Wednesday 1. Sample answer: 1/2, 3/6 2. 20/3 3. < 4. 155/24 bushels

Thursday 1. A (0, 8), B (3, 5), C (4, 7), D (4,2), E (5, 5), F (7, 9), G (8, 4), H (9, 9), I (9, 2)

2. The point should be at (2, 2) 3. The point should be at (0, 3)

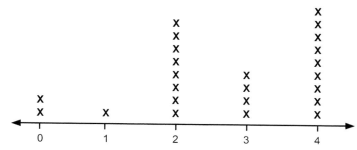

Friday   **1.**

2. Sample answer: Most students visited the library at least twice. Two students did not go to the library at all.

Brain Stretch   10 handshakes

Week 5, pages 13–15

Monday   **1.** B   **2.** 280   **3.** Add 6 if previous number ends with 7 otherwise add 4.   **4.** 56

Tuesday   **1.** The 2 in 742 represents 2 ones. The 2 in 724 represents 2 tens or 20.   **2.** 10^3 **3.** 9.32   **4.** 60.74

Wednesday   **1.** 1/2   **2.** >   **3.** 1/10 lb.   **4.** It is equidistant from 1 and 1/2.

Thursday   **1.** A   **2.** Accept any two closed figures with straight sides.   **3.** quadrilateral   **4.** A, C

Friday   **1.** 1,200 months   **2.** 3,520 yards   **3.** 272 cm perimeter and 4,480 cm² area   **4.** 5

Brain Stretch   7:40 a.m.

Week 6, pages 16–18

Monday   **1.** 182   **2.** 8, 18, 11, 15, 9   **3.** C, C   **4.** 2 × (2 + 12) ÷ (4 + 3) = 4

Tuesday   **1. a)** 34 **b)** 340 **c)** 3,400 **d)** Sample answer: For each zero in the power of 10, there is one zero in the product.

   2. 10^5 **3.** 7   **4.** <

Wednesday   **1.** Sample answer: 8/14, 12/21   **2.** <   **3.** 25/6   **4.** 40/7 lb., It lies between 5 and 6.

Thursday   **1.** (1, 3), (2, 7), (3, 5), (4, 2), (6, 8), (7, 9), (7, 4), (9, 8), (9, 3)   **2.** A point should be drawn at (0, 0).

   3. A point should be drawn at (8, 1).

Friday   **1.** 368 ounces   **2.** 10,000 pennies   **3.** 16,800 m   **4.** 9 cubic units

Brain Stretch $76

Week 7, pages 19–21

Monday   **1.** B   **2.** 68, 59, 95, 104, 41   **3.** 164   **4.** =

Tuesday   **1.** 900 ones **2.** 186.302 = 100 + 80 + 6 + 0.3 + 0.002   **3.** =   **4.** 55

Wednesday   **1.** 7/9   **2.** <   **3.** 1/18   **4.** 3 3/4 cups

Thursday   **1.**

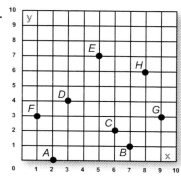

   2. A   **3.** E

Friday   **1.** 10   **2.** 5   **3.** Chris   **4.** They planted the same

Brain Stretch   480 m³

Week 8, pages 22–24

Monday   **1.** D   **2.** Output = Input × 11; 66, 99   **3.** =   **4.** 75 − 25

Tuesday   **1.** 800 hundreds   **2.** 16,000   **3.** 0.06   **4.** 2,700

Wednesday   **1.** Sample answer: 5/11, 20/44 **2.** 41/8   **3.** =   **4.** 1/6

Thursday   **1.** A: 6 sides, B: 3 sides, C: 5 sides, D: 4 sides, E: 4 sides, F: 8 sides, G: 3 sides, H: 4 sides, I: 4 sides, J: 4 sides   **2.** Quadrilaterals: D, E, J; Not Quadrilaterals: A, B, C, F, G, H and I.

Friday   **1.** 45 minutes   **2.** 3,600 m, 18,000   **3.** 34.6 sq yd.   **4.** 20 cubic units

Brain Stretch   He does not have enough money. Both sweaters cost $95.

Week 9, pages 25–27

Monday   **1.** B   **2.** 41   **3.** 3, larger   **4.** (12 + 8) ÷ 2

Tuesday   **1.** 40 tens   **2. a)** 1/10 of the 4 to the left **b)** 10 times the 4 to the right   **3.** <
    4. 872.549 = 800 + 70 + 2 + 0.5 + 0.04 + 0.009

Wednesday   **1.** 2 6/9   **2.** 29/30   **3.** <   **4.** 9 cups

Thursday   **1.** B   **2.** Both have 4 sides and 2 pairs of parallel sides.   **3.** B, C   **4.** congruent

Friday   **1.** 204 inches   **2.** 112.5 words   **3.** Room A has larger area and perimeter.   **4. a)** 7 **b)** 28

Brain Stretch   3 1/6 pints

Week 10, pages 28–30

Monday   **1.** 223   **2.** 90, 120, 10, 30, 60   **3.** (6 + 5) × 3   **4.** D

Tuesday   **1.** 100 ten thousands   **2. a)** 300 **b)** 3000 **c)** 10   **3.** 670,000   **4.** 7.1

Wednesday   **1.** 13/45   **2.** <   **3.** 1/20   **4.** 14 inches

Thursday   **1.**

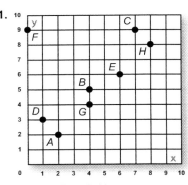

2. *B, G; C, F*   **3.** None

Friday   **1.** Chocolate 5 pints, Chocolate chip 9 pints, Strawberry 10 1/2 pints, Vanilla 6 1/2 pints

    2. Chocolate 11 pints left, Chocolate chip 7 pints left, Strawberry 5 1/2 pints left, Vanilla 9 1/2 pints left

Brain Stretch   **a)** 512 small boxes   **b)** 3,072 buttons

Week 11, pages 31–33

Monday   **1.** 25   **2.** Output = 100 × Input; 800, 1000, 11   **3.** (8 + 56) ÷ 4   **4.** (16 + 4) × (8 − 1) = 140

Tuesday   **1.** 20,000 ones   **2.** 5,100,000   **3.** <   **4.** 5,856

Wednesday   **1.** Sample answer: 6/10, 9/15   **2.** 1 1/13   **3.** 3 ft.   **4.** She does not have enough. 89/6 yards

Thursday   **1.** A   **2.** A, D   **3.** 2   **4.** Similar

Friday   **1.** 2,300 mL   **2.** 235,000 cm²   **3.** C   **4.** 70 sq ft.

Brain Stretch   Honey Hoops is cheaper. Fruity Flakes is $2.75/500 g

Week 12, pages 34–36

Monday   **1.** A   **2.** 55   **3.** 11 − 5 +2   **4.** 16 + 3 × 4 − (2 + 5) = 21

Tuesday   **1.** 12.85   **2.** <   **3.** 63.6   **4.** 44

Wednesday   **1.** 9.5   **2.** 1/35   **3.** 0   **4.** 8/3 servings

Thursday   **1.** Quadrilaterals: B, C, H, E; Parallelograms: B, C, E   **2.** A, D, F, G

Friday   **1.** 1,460   **2.** 10 cm   **3. a)** 4 **b)** 40   **4. a)** 3 × 2 = 6 in.² **b)** 30 in.³

Brain Stretch   Yes he can. He has 7/8 gallons in total.

Week 13, pages 37–39

Monday   **1.** 1, 3, 5, 7, 9; 1, 4, 7, 10, 13 **b)** Sample answer: The difference between corresponding terms starts at 0 and increases by 1 each time; 1 is the difference between the number being added in each pattern rule.

2. 226   **3.** *b* + 11

Tuesday   **1.** 300 hundreds   **2.** 7,373,082   **3.** >   **4.** 31.5

Wednesday   **1.** 13/14   **2.** 14/5   **3.** 4   **4.** 7/4 of a sandwich.

Thursday   **1.**

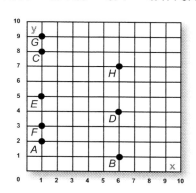

2. All points are on the same vertical line, at 1 on the *x*-axis.

3. All points are on the same vertical line, at 6 on the *x*-axis.

Friday   **1.** 27 feet   **2.** 200 dimes   **3.** 4000/19 m/min   **4. a)** 12 cu in. **b)** 36 cu in.

Brain Stretch   45/4, $11.25

Week 14, pages 40–42

Monday   **1. a)** 1, 3, 5, 7, 9; 1, 4, 7, 10, 13 **b)** (1, 1), (3, 4), (5, 7), (7, 10), (9, 13)

c) The points lie on a straight line.

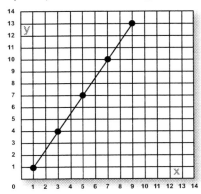

Tuesday   **1.** 94.57   **2.** hundredths   **3.** =   **4.** 4.32   **5.** 5,829.429

Wednesday   **1.** 19/14   **2.** 23/4   **3.** >   **4.** 28/5 hours a week, 224/5 hours every 8 weeks

Thursday   **1.** B, C, D   **2.** A, B   **3.** The closed shape should have 5 straight sides of different lengths.

4. Sample answer: acute triangle

Friday   **1.** 175, 100, 125, 225   **2.** Saturday   **3.** 100 to 225 tickets were sold.   **4.** 100 tickets increase

5. Sample answer: Tickets Sold This Week

Brain Stretch   2/3 cups of sugar

Week 15, pages 43–45

Monday   **1.** 5   **2.** (5 + 7) × (2 + 3)   **3. a)** C **b)** P   **4.** 1 and 7

Tuesday   **1.** 536,000   **2.** hundredths   **3.** 30   **4.** 18.62

Wednesday   **1.** 2 1/8   **2.** 1/4   **3.** 3/5   **4.** 3.25 L

Thursday   **1. a)** and **b)** Rectangle   **2. a)** and **b)** (3, 5), (5, 8), and (6, 6)

Friday   **1.** 1,300 cm   **2.** $4.05/gallon   **3.** Perimeter: 36 m; Area: 81 m²

4. a) Length: 4 cm; Height: 2 cm; Width: 2 cm **b)** 16 cm³

Brain Stretch   8:23 a.m.

Week 16, pages 46–48

Monday   **1. a)** 12, 11, 10, 9, 8; 12, 9, 6, 3, 0 **b)** The difference between corresponding terms starts at 0 and increases

by 2 each time; 2 is the difference between the numbers being added in each pattern rule.

2. 14   **3.** 16 ÷ 8 + 43

Tuesday   **1.** 4,000 tens   **2.** 7   **3.** <   **4.** 675.398 = 600 + 70 + 5 + 0.3 + 0.09 + 0.008

Wednesday   **1.** 59/4   **2.** <   **3.** 1/24   **4.** 4 tablespoons

Thursday   **1.**

2. no parallel sides: B, C, E; 1 pair of parallel sides: F; 2 pairs of parallel sides: A, D, G, H

Friday   **1.** 1 X should be added at 2, 3 Xs added at 4, and 3 Xs added at 4 1/2.   **2.** 47 nails   **3.** Yes
4. No, he needs 5 more 5-inch nails.

Brain Stretch   5 in. × 6 in.; 10 in. × 3 in.; 1 in. × 30 in.

Week 17, pages 49–51

Monday   **1. a)** 12, 11, 10, 9, 8; 12, 9, 6, 3, 0 **b)** (12, 12), (11, 9), (10, 6), (9, 3), (8, 0)
c) All points lie on a straight diagonal line.

Tuesday   **1.** 60,000 ones   **2. a}** 0.0216 **b)** Sample answer: The decimal point moved 4 places to the left in the quotient.
3. <   **4.** 98

Wednesday   **1.** 3 1/4   **2.** 17/6   **3.** 3 × **1**/3 = 1   **4.** 6 students

Thursday   **1.** (1, 7), (5, 7), (5, 5), (1, 5)   **2.** 4 units; 2 units   **3.** Sample answer: M (1, 7), N (9, 7), O (9, 3), P (1, 3)

Friday   **1.** 9,000 g   **2.** 100 nickels   **3.** George is farther by 114 cm.   **4.** 5 m^3

Brain Stretch   Sample answer: 6 dimes and 4 pennies; 1 quarter, 3 dimes, 1 nickel and 4 pennies; 7

Week 18, pages 52–54

Monday   **1.** (50 − 40) × 8   **2.** 11   **3.** 2, smaller   **4.** 1, 2, 2, 2, 2, 3, 48

Tuesday   **1.** 235,000   **2.** 13.3   **3. a)** 3,500 **b)** Sample answer: The decimal point moved 3 places to the right in
the product.   **4.** 9.852   **5.** 420.679

Wednesday   **1.** 1   **2.** 1/2   **3.** 20   **4.** 31 1/2

Thursday   **1.** B, D   **2.** C   **3.** Accept any two quadrilaterals.   **4.** Similar

Friday   **1.** 10 cups   **2.** 21 liters   **3.** Perimeter: 32 yd.; Area: 28 sq yd.   **4.** 48 cu in.

Brain Stretch   No he spent 3/5 on the video game.

Week 19, pages 55–57

Monday   **1. a)** 4, 6, 8, 10, 12; 0, 2, 4, 6, 8 **b)** The difference between corresponding terms is the same, 4
2. (3 + 16) × 5   **3.** 100, larger

Tuesday   **1.** 70 ten thousands   **2.** 3,000,000   **3.** 2   **4.** 7.04

Wednesday   **1.** 1 5/12   **2.** 11 1/2   **3.** 133/6 m^2 **4.** $1

Thursday   **1.** Bakery (2, 2); Bank (11, 8); Flower shop (7, 6); Gas station (6, 1); Pet store (12, 5); Restaurant (3, 9)

Friday   **1.** 22   **2.** $44   **3.** 0

Brain Stretch   **a)** length = 3w + 8 **b)** perimeter = 8w + 16

Week 20, pages 58–60

Monday   **1. a)** 4, 6, 8, 10, 12; 0, 2, 4, 6, 8 **b)** (4, 0), (6, 2), (8, 4), (10, 6), (12, 8)
c) All points lie on a straight diagonal line.

Tuesday   **1.** 57   **2.** 8   **3.** 61.47   **4.** 6,720

Wednesday   **1.** 98/11   **2.** 87/7   **3.** 0   **4.** 1/8

Thursday   **1.** A, B, C, D   **2.** trapezoid   **3.** B, C, D   **4.** 5

Friday   **1.** 6,700 mL   **2. a)** Length 3 inches; Width 3 inches; Height 4 inches **b)** 36 cu in.
3. 40 sq ft.   **4.** 12 sq ft.

Brain Stretch   **a)** 30 cups **b)** No, you would need 60 cups.

Week 21, pages 61–63

Monday   **1.** 13   **2.** (11 – 5) × 2   **3.** 5 larger   **4.** 7, 14, 21, 28, 35

Tuesday   **1.** 200 tens   **2.** 64.36   **3.** 124   **4.** 136

Wednesday   **1.** 24/7   **2.** 27/4   **3.** 8/5 sq yd.   **4.** 4 1/2

Thursday   **1.** and **2.**

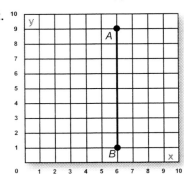

    3. Sample answer: (6,1), (8,1)   **4.** Sample answer: (4, 1), (4, 9)

Friday   **1. a)** 3,600 seconds **b)** 480 minutes **c)** 17 hours   **2.** 10 m³   **3.** 18 2/3 feet long

Brain Stretch   **a)** 0.60   **b)** 8 calls

Week 22, pages 64–66

Monday   **1. a)** 1, 3, 5, 7, 9; 3, 4, 5, 6, 7 **b)** The first sequence is growing more quickly than the second sequence but starts at a lower number.   **2.** a + 50   **3.** C, C

Tuesday   **1.** 700 ones   **2.** 34   **3.** 700   **4.** 94.83

Wednesday   **1.** 37/12   **2.** 1/72   **3.** 12/7   **4.** 59/12 pints

Thursday   **1.** Bison (5, 7); Crocodiles (3, 3); Elephants (12, 9); Gorillas (12, 1); Hippopotamuses (2, 8); Leopards (10, 4)

Friday   **1.** Shading for should go to 50 for Class A, 43 for Class B, 63 for Class C, 38 for Class D, and 57 for Class E.

    2. Class C   **3.** 19 students   **4.** 251 books   **5.** Classes A, C, D, E

Brain Stretch   11 squares

Week 23, pages 67–69

Monday   **a)** 1, 3, 5, 7, 9; 3, 4, 5, 6, 7 **b)** (1, 3), (3, 4), (5, 5), (7, 6), (9, 7)   **c)** A straight diagonal line.

Tuesday   **1.** thousands **2.** 700,000 + 70,000 + 8,000 + 100 + 40 + 5 + 0.2 + 0.09   **3.** 84.12   **4.** 3,220

Wednesday   **1.** 1 4/5   **2.** 3/3   **3.** 4/9   **4.** 14 stuffed animals left

Thursday   **1.** A, C   **2.** Accept any 3 correctly labeled triangles.   **3.** A   **4.** T

Friday   **1.** 33,520 mL   **2.** 3 ft.   **3.** Perimeter: 360 in., Area: 7,200 sq in.; No, it is much too small.

    4. 7 hours and 25 minutes

Brain Stretch   6 1/2 cups of smoothie; 2 1/6 cups per serving

Week 24, pages 70–72

Monday   **1.** 15   **2.** 97 – m   **3.** 24   **4.** 1, 2, 17, 34

Tuesday   **1.** 1,000 thousands   **2.** 0.32   **3.** 21.1   **4.** 30.36

Wednesday   **1.** 47/6   **2.** <   **3.** 36   **4.** 1/8 gallons

Thursday   1.

2. 0 right angles: C, D, E; 1 right angle: A and H; 2 right angles: G; 3 right angles: none; 4 right angles: B, F

Friday   1.

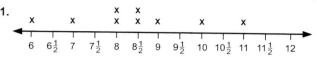

2. 76 1/2 L   3. 19; 1/2 L   4. 2 1/9 cans per family

Brain Stretch   **a)** 3 **b)** $1,267.97

Week 25, pages 73–75

Monday   **1.** 70 ÷ (5 + 2)   **2.** 57   **3.** 4, larger   **4.** 1 and 2

Tuesday   **1.** 220,000   **2.** eight hundred seventy two and five hundred forty-nine thousandths

3. thousandths   **4.** 4,042   **5.** 752.862

Wednesday   **1.** 73/8   **2.** 1/7   **3.** 28/3 miles   **4.** 3/5 hour at the dinosaur exhibit and 12/5 hours at the moon exhibit

Thursday   **a)** Archery (5, 5); Cycling (9, 2); Equestrian Jumping (7, 5); Gymnastics (2, 8); Soccer (3, 3);

Weightlifting (9, 10) **b)** Equestrian Jumping

Friday   **1.** 1,200 seconds   **2.** No, to give 25 people 2 oz. each, Samantha would need 3 pounds of jelly beans and

she only has 32 oz.   **3.** Perimeter: 2,736 in.; Area: 466,560 sq in.   **4.** Area: 104 m²; Volume: 60 m³

Brain Stretch $44.99

Week 26, pages 76–78

Monday   **1.** (100 − 10) ÷ 9   **2.** 20   **3.** 3, larger   **4.** 1,11

Tuesday   **1.** 0.5   **2.** 4   **3.** 5   **4.** 112

Wednesday   **1.** 4 5/11   **2.** 35/64   **3.** 20   **4.** 29 1/2 gallons

Thursday   **1.** B, C, D   **2.** B, E   **3.** 1   **4.**

Friday   **1.** 1.2 kg   **2.** 180 cu ft.   **3.** 528 cu in.   **4. a)** 24 m **b)** 27 m²

Brain Stretch   **a)** 11 pounds **b)** 7 complete pies

Week 27, pages 79–81

Monday   **a)** 0, 3, 6, 9, 12; 0, 9, 18, 27, 36 **b)** The terms in the second pattern are 3 times the corresponding terms in

the first pattern; the number added in the first pattern rule (9) is 3 times the number added in the second

pattern rule (3). **c)** (0, 0), (3, 9), (6, 18), (9, 27), (12, 36) **d)** A straight diagonal line

Tuesday   **1.** 8,500   **2.** 30,000   **3.** 65.25   **4.** 55   **5.** 274.386

Wednesday   **1.** 11 2/5   **2.** 54   **3.** 1/25 sq yd.   **4.** 4 1/2 rolls

Thursday   **1.** B   **2.** No, for an obtuse angle there must be a long side and a short side.   **3.** 1   **4.** True

Friday   **1.** 93, 160, 70   **2.** 18.6, 32, 14   **3.** Week 2; Sample answer: There was enough time for people to hear

about the food drive.   **4.** More cans are donated in the middle of the week; Sample answer: People are

busier on Mondays and Fridays.

Brain Stretch   19 jars and 4 cartons

Week 28, pages 82–84

Monday   **1.** P, C   **2.** $120 - y$ **3.** <   **4.** number of flowers/flowers per bunch = number of bunches; 8 bunches

Tuesday   **1.** 420 + 21 = 441   **2.** Sample answer: The 2 in 345.42 is 1/10 of the value of the 2 in 743.25.

                     3. 30,250   **4.** 484

Wednesday   **1.** 6   **2.** 3/5   **3.** 22.5 miles   **4.** 7 1/2 servings

Thursday   **1.** Rectangles will vary.   **2.** Drawings should show one vertical and one horizontal like through the center.

Friday   **1.** 15 cu ft.   **2. a)** 940 g **b)** No

Brain Stretch   37 pizzas

Week 29, pages 85–87

Monday   **1.** (1, 3)   **2.** =   **3.** 2   **4.** 22 + 8

Tuesday   **1.** 0.539   **2.** 63,000,000   **3.** $43.47   **4.** 3,128

Wednesday   **1.** Sample answer: less than; 8 is multiplied by a factor less than 1, so the product must be less than 8.

                     2. 9/4   **3.** =   **4.** 5/24 quart

Thursday   **1.** Square and rhombus both had all equal sides. Rectangles, squares and rhombuses have 4 parallel sides. Rhombuses have non right angles which is the difference.

                     2. No, the angles inside a triangle must add to 180°.   **3.**   **4.** Congruent

Friday   **1.** 5,900 grams   **2.** 1 hour and 29 minutes   **3.** 1,458 cu in.   **4.** Area: 160 sq ft.; Volume: 96 cu ft.

Brain Stretch   41 students voted

Week 30, pages 88–90

Monday   **a)** 0, 8, 16, 24, 32; 0, 4, 8, 12, 16 **b)** Each term in the second pattern is half the corresponding term in the first pattern; the number being added in the second pattern rule, 4, is half the number being added in the first pattern rule. **c)** (0, 0), (8, 4), (16, 8), (24, 12), (32, 16) **d)** A line

Tuesday   **1.** 60,000   **2.** 10   **3.** 17,045   **4.** 2.29

Wednesday   **1.** Sample answer: greater than; 2 1/4 × 7 is greater than 7 because 2 groups of 7 is 14 and 2 1/4 is a little more than 2 groups of 7.   **2.** 1.76 m²   **3.** 18   **4.** 49/18 days

Thursday   Sample Answer: 3 Sides: E, G, D; 4 Sides: A, B, I, K; 1 Side: C; More than 4 Sides: F, H, J

Friday   **1.** They can use the two 6-ft. planks as 6-ft. planks. They can cut four more 6-ft. planks from all the 10-ft. planks, and cut four 3 1/2-ft. planks from the 4-ft. leftover pieces. There are four 1/2-ft. pieces of wood leftover.

                     2. Sample answer: use the 6-ft. planks; cut two more 6-ft. planks from the 7-ft. planks; cut two 3 1/2-ft. planks from each of two 8-ft. planks.

Brain Stretch   Estimate the shape of the puddle with short straight lines and measure them all and add them all to get the perimeter. The shorter the lines, the more accurate the answer will be.

Common Core State Standards for Mathematics Grade 5

Student	5.OA.1	5.OA.2	5.OA.3	5.NBT.1	5.NBT.2	5.NBT.3	5.NBT.4	5.NBT.5	5.NBT.6	5.NBT.7	5.NF.1	5.NF.2	5.NF.3	5.NF.4	5.NF.5	5.NF.6	5.NF.7	5.G.1	5.G.2	5.G.3	5.G.4	5.MD.1	5.MD.2	5.MD.3	5.MD.4	5.MD.5

Level 1 Student demonstrates limited comprehension of the math concept when applying math skills.

Level 2 Student demonstrates adequate comprehension of the math concept when applying math skills.

Level 3 Student demonstrates proficient comprehension of the math concept when applying math skills.

Level 4 Student demonstrates thorough comprehension of the math concept when applying math skills.

Math — Show What You Know!

_____

☐ I read the question and I know what I need to find.

☐ I drew a picture or a diagram to help solve the question.

☐ I showed all the steps in solving the question.

☐ I used math language to explain my thinking.